Facts, Conjectures, and Improvements for Simulated Annealing

SIAM Monographs on Mathematical Modeling and Computation

About the Series

In 1997, SIAM began a new series on mathematical modeling and computation. Books in the series develop a focused topic from its genesis to the current state of the art; these books

- present modern mathematical developments with direct applications in science and engineering;
- describe mathematical issues arising in modern applications;
- develop mathematical models of topical physical, chemical, or biological systems;
- present new and efficient computational tools and techniques that have direct applications in science and engineering; and
- illustrate the continuing, integrated roles of mathematical, scientific, and computational investigation.

Although sophisticated ideas are presented, the writing style is popular rather than formal. Texts are intended to be read by audiences with little more than a bachelor's degree in mathematics or engineering. Thus, they are suitable for use in graduate mathematics, science, and engineering courses.

By design, the material is multidisciplinary. As such, we hope to foster cooperation and collaboration between mathematicians, computer scientists, engineers, and scientists. This is a difficult task because different terminology is used for the same concept in different disciplines. Nevertheless, we believe we have been successful and hope that you enjoy the texts in the series.

Joseph E. Flaherty

Peter Salamon, Paolo Sibani, and Richard Frost, *Facts, Conjectures, and Improvements for Simulated Annealing*

Lyn C. Thomas, David B. Edelman, and Jonathan N. Crook, *Credit Scoring and Its Applications*

Frank Natterer and Frank Wübbeling, *Mathematical Methods in Image Reconstruction*

Per Christian Hansen, *Rank-Deficient and Discrete Ill-Posed Problems: Numerical Aspects of Linear Inversion*

Michael Griebel, Thomas Dornseifer, and Tilman Neunhoeffer, *Numerical Simulation in Fluid Dynamics: A Practical Introduction*

Khosrow Chadan, David Colton, Lassi Päivärinta, and William Rundell, *An Introduction to Inverse Scattering and Inverse Spectral Problems*

Charles K. Chui, *Wavelets: A Mathematical Tool for Signal Analysis*

Facts, Conjectures, and Improvements for Simulated Annealing

Peter Salamon

San Diego State University
San Diego, California

Paolo Sibani

University of Southern Denmark
Odense, Denmark

Richard Frost

San Diego State University
San Diego, California

siam

Society for Industrial and Applied Mathematics
Philadelphia

10 9 8 7 6 5 4 3 2 1

Library of Congress Cataloging-in-Publication Data

Salamon, Peter, 1950-
Facts, conjectures, and improvements for simulated annealing / Peter Salamon, Paolo Sibani, Richard Frost.
p. cm. — (SIAM monographs on mathematical modeling and computation)
ISBN 0-89871-508-3 (pbk.)
1. Simulated annealing (Mathematics) I. Sibani, Paolo, 1954- II. Frost, Richard, 1955- III. Title. IV. Series.

QA402 .S35 2002
519.3—dc21 2002029215

Contents

List of Figures

Preface

Simulated annealing is a simple and general algorithm for finding global minima. It operates by simulating the cooling of a (usually fictitious) physical system whose possible energies correspond to the values of the objective function being minimized. The analogy works because physical systems occupy only the states with the lowest energy as the temperature is lowered to absolute zero.

Simulated annealing has been developed by a wide and highly interdisciplinary community and used by an even wider one. As a consequence, its techniques and results are scattered through the literature and are not easily accessible for the computer scientist, physicist, or chemist who wants to become familiar with the field.

The present monograph is intended both as an introduction for the noninitiated and as a review for the more expert reader. We start by developing the subject from scratch. We then explain the methods and techniques indispensable for building state-of-the-art implementations. The physical background presented is meant to sharpen the reader's intuition for the field.

In our choices we have been somewhat biased toward the line of investigation developed in the last decade and a half of progress in one particular community: those who have worked on the theory of optimal thermodynamic processes and obtained most of their results by viewing simulated annealing as the cooling of a physical system. This point of view leads to some important improvements in algorithm design and these form the main topic discussed herein. While some proofs for these improvements are presented, most of the results require assumptions that may or may not be justified and should therefore be classified as conjectures. As such, they represent open problems in the area, and the tone of the discussions is in this spirit. Experimental data and heuristic arguments regarding various improvements are presented and discussed. Special care is taken to make the underlying assumptions explicit and to examine the likelihood of their validity.

The background assumed is minimal. The little physical background required is derived in Chapter 5; the reader with a background in physical science can probably skim this part. The mathematical background, mostly the theory of finite Markov chains, is developed in Chapter 6. The reader with an educational background in mathematics or computer science can probably skim section 6.1. The rest of Chapter 6 is required reading for all undergraduate backgrounds and brings together those optional topics from statistical physics and Markov chains that bear strongly on the reader's understanding of the effect that variants of the algorithms can have. Only readers with a graduate background in statistical physics can skim this part without compromising later understanding.

The book is divided into four parts, each with a distinct purpose and flavor. Minimal familiarity with a part can be gained by reading the brief synopsis at its beginning. Part I presents an overview of simulated annealing and how it fits into the family of algorithms for global optimization. This part is accessible to all readers and they should familiarize themselves with it before proceeding. Part II develops the theoretical aspects of complex systems that are indispensable for understanding simulated annealing and its refinements. We tried hard to keep this material to a minimum so as to make Part III of the text as accessible as possible to a wide audience. The further one has managed to work through Part II, the more the refinements in Part III will make sense. We have found it relatively easy to cover Parts I and II in their entirety with advanced undergraduate classes of mixed backgrounds (mathematics, computer science, physics, and engineering) in a one-semester course using selected topics from Part III as a supplement and source of projects. Part III is very different in nature from the previous parts. Theory is almost completely absent and each chapter focuses on an area where simulated annealing can be enhanced and describes the competing methods for improving performance. Each of the methods presented has been tested on some examples on which its creators claim improved performance. It is our belief that selecting the more promising algorithms will depend on the structure of the problem of interest, i.e., different enhancements improve performance on different problems. Some of the enhancements are outlined in sufficient detail to allow implementation; others require going back to the research articles cited for further details. In general, Part III requires significantly more sophistication on the part of the reader. Coding an enhancement based on our description requires thorough understanding. Part IV switches back to the style in Part II, albeit at a significantly more advanced level. It starts off roughly where Part II leaves off and continues our development of the physical understanding that one can get about global optimization problems. This is definitely graduate-level material and brings the reader to the research level in the field. While the theory developed here is not indispensable for utilizing the enhancements in Part III, it does shed further light on these methods and is, in our opinion, indispensable for the ultimate understanding that can bring this subject from the realm of heuristics to the realm of provably optimal algorithms.

The text grew out of a number of courses taught at the Niels Bohr Institute, the Technical University of Denmark, the University of Southern Denmark (formerly Odense University), and San Diego State University. A version of this course also formed the material presented in a tutorial for the SIAM annual meeting during the summer of 1994 in San Diego.

Additional material related to the contents of this book can be found at `http://www.frostconcepts.com/`. Specific questions regarding errata and updates can be addressed to `salamon@math.sdsu.edu`.

Acknowledgments

Thanks to ...

Our families for enduring many days and nights of absence.

Our colleagues and students for their many inspiring questions.

The publishers for their incredible patience with repeated delays in the writing process.

Jacob Mørch Pedersen, from whose Ph.D. dissertation we have borrowed numerous results.

Lene Mørch Pedersen for consenting to our use of her paper silhouette in Figure 13.1.

Gale Bamber of the San Diego Supercomputer Center for her graphic interpretation of energy landscapes displayed on the cover.

Tove Nyberg of SDU-Odense Universitet for drawing most of the figures.

Statens Naturvidenskabelige Forskningsråd for financial support.

Part I

Overview

Part I of this book presents a stand-alone course on the very elementary aspects of annealing. The really impatient reader can go directly to Chapter 4, which presents the bare-bones annealing algorithm. On the other hand, the experienced reader may choose to skip Chapter 4.

Chapter 1 puts the annealing algorithm in context and compares and contrasts the annealing approach to algorithms based on biological heuristics. This chapter is strongly recommended for all readers. Chapter 2 presents six examples of simulated annealing problems. These problems are used throughout the text for illustration. The impatient reader can come back to these as needed. Chapter 3 consists of an elaborate table to be used as a dictionary for translating different terms used in the nomenclature of global optimization and its physical and biological counterparts. A glance at this table is recommended and it can be consulted as needed.

Chapter 1
The Place of Simulated Annealing in the Arsenal of Global Optimization

This text presents one general approach to the problem of global optimization. The basic problem is to find the lowest possible value of a function[1] whose graph looks typically like a many-dimensional version of Fig. 1.1. Six examples of such problems are described in Chapter 2.

Calculus provides a powerful tool for characterizing and locating *local* minima. The problem of *global* minima has remained much more elusive. In the special case of a convex function,[2] the familiar conditions from calculus are necessary and sufficient conditions with which to establish the optimality of a point. Lacking convexity, one is left with little recourse except to enumerate all possibilities and examine each one (sometimes referred to as a "grand tour"). This is not a viable option for the type of problems we are interested in, which typically have far too many candidates for exhaustive enumeration.[3] Lacking a better strategy, enumeration until the point of exhaustion does give a viable strategy for finding better than average solutions even if the very best is unreachable. Computers have had a large impact here since they have extended the point of exhaustion by many orders of magnitude. In fact, in the 1950s the computer was heralded as a wonderful tool for global optimization because it made long random search runs possible [Con80]. A method often advocated was to evaluate the function at a few million points and select the best value found. A variant of this scheme is a good strategy in low dimensions or in problems with relatively few minima: use many random points as initial states and feed them to a local minimizer.[4]

Random search was not the last word in global optimization, but it did bring to the field the important tool of a random number generator. Methods that incorporate the use of random number generators are known as Monte Carlo methods. Typically such methods

[1] The problem of finding the largest possible value of a function f is equivalent to finding the lowest possible value of $-f$.

[2] A function is convex iff a line segment connecting two points on the graph of the function is always above the graph of the function [Lue84].

[3] Berry et al. [BBK+96], however, carried out such a program for chemical structure problems in clusters of moderate size (see Example B in Chapter 2).

[4] A more sophisticated variant uses simulated annealing combined with a local minimizer [DW96, MRX92]. This technique is called basin hopping and is discussed in Chapter 10.

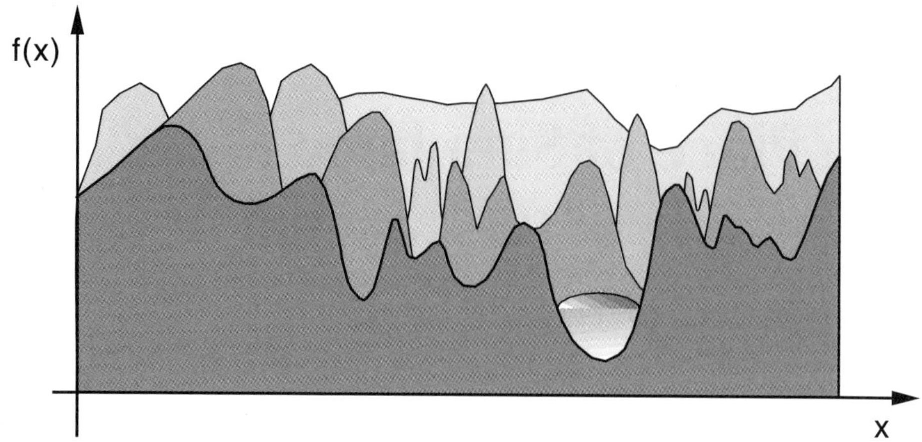

Figure 1.1. *A landscape with many local minima.*

are superior to deterministic algorithms for problems in high dimensions (for example, in numerical evaluation of integrals). Clever implementations of Monte Carlo techniques amount to refinements of random search known as importance sampling [BH97]. Such sampling techniques can greatly increase the rate of convergence of Monte Carlo algorithms by sampling preferentially according to some distribution. Choosing the best distribution for rapid convergence is often an art [BH97]. We will find that this idea of *importance sampling* is very useful for comparing simulated annealing with other Monte Carlo–based global optimization procedures. The differences between the techniques are nicely characterized by differences in their criteria of importance during the sampling.

During the last 15 years, Monte Carlo methods have taken over the field of global optimization in the form of heuristic approaches based on analogies [Hol75, Aarts89, Otten89, Gold89, HRS98, Koza99]. The algorithms set up a correspondence between the optimization problem and a physical or biological system. These algorithms simulate the behavior of the system to find states with good values of the objective function. While myriad variations on both themes exist, the main choices for the correspondence are either the slow cooling of a physical system to find low-energy states or the evolution of a biological population to find high-fitness states. The first is called simulated annealing and is the basis of physical heuristics. The second is called either genetic algorithms or evolutionary programming and is the basis of biological heuristics. In the context of importance sampling, the major difference between the two schemes can be understood as a degree of greediness or a degree of urgency.

An optimization algorithm is called *greedy* if it chooses the best alternative at each step without regard to long-term benefits. The standard example is the method of steepest descent [Lue84]. The relevance for optimization heuristics is as follows. All heuristics in the family under discussion search the state space by a sequence of random choices of states. They all sample preferentially "near" states with good values of the objective. This is in keeping with the dictates of importance sampling. Biological algorithms tend to add a bias in favor of those states with good values of the objective that were found quickly. That

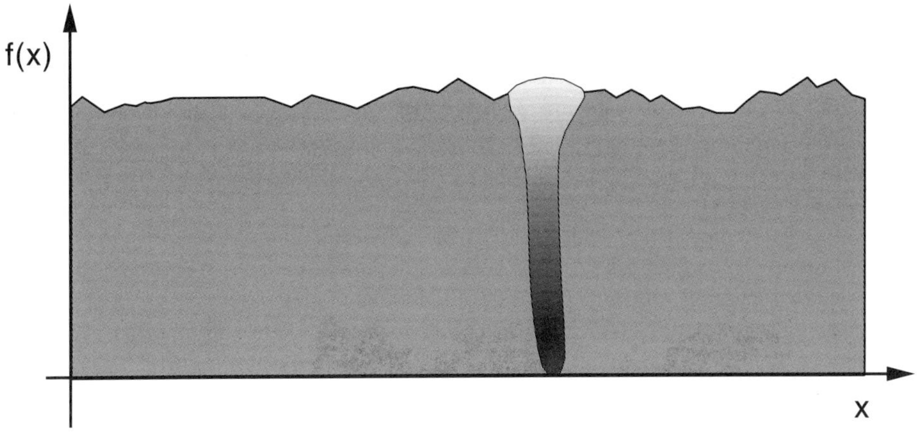

Figure 1.2. *A golf-hole landscape.*

is, they preferentially "breed" those members of the population (states) with high fitness. The physical heuristics on the other hand search to the same extent near all states at a given value of the objective. To readers with a background in biochemistry, this distinction may appear familiar as the one of thermodynamic versus kinetic control of a reaction. In the language of importance sampling, the issue is what criterion constitutes importance. The physical heuristic places importance only on the quality of the objective and not on how quickly it is reached. The biological heuristics place importance also on how quickly the high-fitness state is reached. In this sense, the biological algorithms are more greedy.

How the performances of various heuristic methods compare with each other as well as with more traditional local methods such as quasi-Newton or steepest descents depends on the structure of the problem. For example, if we are dealing with a convex function in n real variables, then calculus-based tools (quasi-Newton, conjugate gradient, etc.) are the best choices. At the opposite extreme, the values of the objective function are randomly assigned to states, with zero correlation between nearness of the points and values of the objective. In this case, sampling near states with good values of the objective does not help, and we might as well use random search. Similarly, no algorithm can beat random search on the so-called golf-hole problem, a variant of which is shown in Fig. 1.2. Such problems are characterized by the fact that there is no information regarding the global optimum at points outside a small neighborhood of this optimum. Between these extremes it is not known which structures lead to which purportedly optimal choices of algorithms beyond the usually exaggerated claims of the inventors. The results depend heavily on who runs the experiments and what examples they choose.[5] One of the themes of the present book is the importance of finding a classification scheme for hard optimization problems. In lieu of firm principles, we are left only with the guidance afforded by the old metatheorem valid for all applications of mathematics: The more we exploit the structure of the problem, the better.

One conclusion based on this metatheorem is that further study of the structure of a problem enables one to design better, more specific algorithms. An example of how

[5] See, however, the experiments described by Johnson et al. [JAMS89, JAMS91] which achieve a higher level of thoroughness.

this often happens for global optimization problems is to identify a well-designed crossover move that blends two known good states in the hopes of creating a better one. Such crossover moves can then be exploited by a genetic algorithm. While a well-designed crossover move that takes advantage of structure can dramatically improve performance, random selection of crossover moves that disregards structure generally does poorly. In terms of our earlier analogy of kinetic versus thermodynamic control, the design of a good crossover move corresponds to the design of a good catalyst; it can significantly lower barriers between local minima and thereby speed up the transition between them.

Simulated annealing assumes and uses only minimal structure. This keeps it simple and general in purpose and is likely to ensure simulated annealing a permanent niche in the arsenal of optimization algorithms. Simplicity takes on additional importance when we note the following fact: The major development in optimization of the past century is the recognition of the overwhelming importance of constraints in industrially useful problems. The special case of a linear objective and linear constraints (linear programming) represents a multibillion dollar boon to today's industries.[6] Similar opportunities are being exploited today for combinatorial optimization problems. Solving a traveling salesman problem to optimize a delivery route has to take into account such exotic constraints as limitations on the weight of the truck when crossing a certain bridge along the route.[7] Each such constraint alters the structure of the problem significantly and provides a strong incentive for sticking with simple, general purpose algorithms.

In summary, simulated annealing is simple, general purpose, and maximally conservative on the kinetic versus thermodynamic control scale. To determine whether simulated annealing is the right tool, we would recommend two things:

1. Determine whether local minima are a problem. Typically this involves greedy searches from a random sample of initial states. If the results are widely disparate, then it probably pays to do a careful search for global as opposed to just local minima.

2. Determine whether the problem warrants exploiting additional structure. This often amounts to the issue of what additional information already exists concerning the problem and algorithms for its solution and how easily this additional information can be implemented in a global optimization algorithm. How some aspects of the structure of a problem can be estimated adaptively is an important theme of this book.

It is in the intermediate regime where simulated annealing finds its niche—complex problems whose structure does not warrant further study or that are so hard that analysis is likely to be very slow. Although we argued that one of the most appealing features of simulated annealing is its simplicity, there are a number of pitfalls to avoid and enhancements to take advantage of. In short, it still pays to understand how simulated annealing works and to take advantage of this understanding for deciding how to best implement the algorithm. The main purpose of this text is to present these pitfalls and enhancements in a generally accessible form.

[6]In 1954 Mobil Oil, in a controversial move, spent several million dollars on a computer system. They recouped their expenses in two weeks of allocating resources using linear programming [Fra83].

[7]We owe this example to Ingo Morgenstern, who used it in a lecture introducing the global optimization workshop at the Telluride Summer Research Center in 1997.

Chapter 2

Six Simulated Annealing Problems

This chapter defines six sample problems to be used in the rest of the book. These problems serve to keep the discussion less abstract by illustrating the various concepts in concrete terms and to present experimental findings of relevance to the discussions. Our problems are toy versions of real applications of industrial significance. Since this is not their main role here, we include only a brief indication of the industrially important problems they represent.

The problems are

A. Seismic deconvolution,

B. Chemical clusters,

C. Spin glasses,

D. Graph bipartitioning,

E. Traveling salesman, and

F. Protein folding.

Some of these problems (D, E) are classic optimization problems in which the global minimum (or the best approximation we can find) is of primary interest. The others are sometimes called sampling problems in which all near-optimal possibilities are of interest.

2.1 Problem Definitions

Problem A: Seismic deconvolution. Geophysical problems often involve so-called inverse problems [JMS96], where one attempts to identify the best theoretical model based on available experimental data. As shown below, this induction process typically requires minimizing a measure of the deviation between model predictions and observed data, over the space of parameter values characterizing the model. Simulated annealing, along with

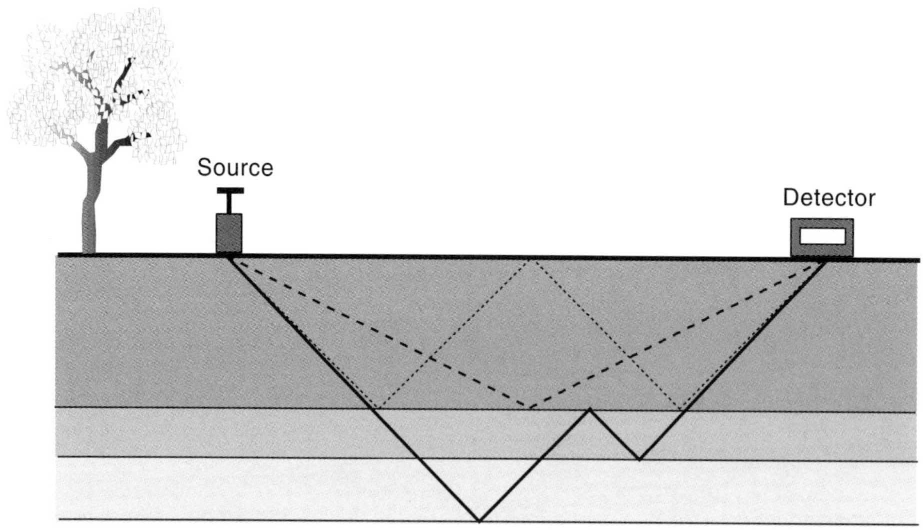

Figure 2.1. *The schematics of an experimental set-up for seismic wave generation and detection. Waves generated at the source are partly reflected at each of the interfaces between layers of different density. The signal reaching the detector is a superposition of these reflections, each reflection delayed by an amount related to the distance traveled. By decomposing the signal, information is obtained about the layers below the ground.*

the enhancements described in this book, has been extensively used by the geophysical community. The model problem considered here describes the detonation of a small explosive charge at a source located on the surface of the earth (see Fig. 2.1) and the interpretation of the echoes of this explosion received at the detector. We assume for simplicity that the subsurface consists of horizontal strata of uniform thickness and density. The sound waves are partially reflected and partially transmitted at each interface, resulting in a detected signal that is a superposition of time translated and attenuated (scaled) copies of the source signal.

Let $f(t)$ be the signal at the source and introduce parameters τ_i and α_i for the delay and attenuation of the ith component in the superposition. The predicted signal at the detector is

$$s_{\text{pred}}(t) = \sum_i \alpha_i f(t - \tau_i). \tag{2.1}$$

The identification problem then reduces to finding the values of the τ_i and α_i, $i = 1, \dots, n$, which minimize

$$\text{Goodness of Fit} = \textstyle\sum_j (s_{\text{obs}}(t_j) - s_{\text{pred}}(t_j))^2, \tag{2.2}$$

where t_j are the sampling times of the detector. This procedure is illustrated in Fig. 2.2.

Real versions of the problem allow for more complicated subsurface structures (tilted strata, pockets, etc.) and additional information regarding echoes collected at nearby locations to create a coherent model of subsurface features over a large region. One clearly

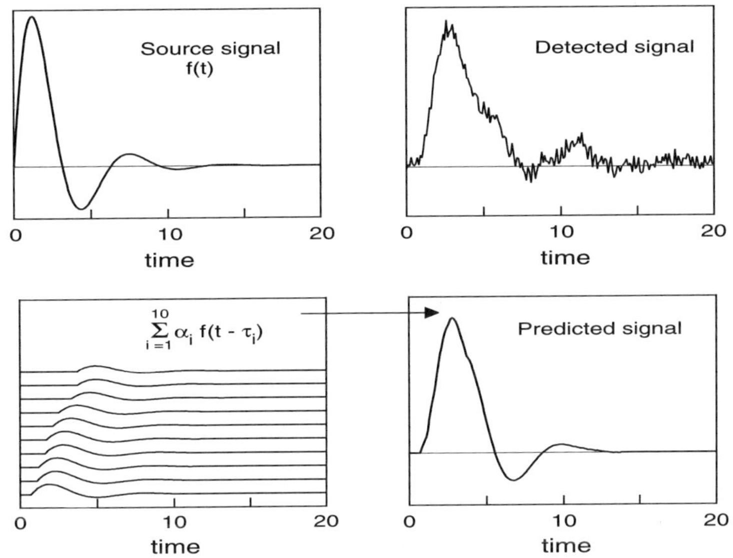

Figure 2.2. *Source, detected, and predicted signals in a seismic wave experiment. In this case, the predicted signal is a sum of shifted and attenuated reflections from various underground layers.*

important application involves geophysical prospecting. A typical result of professional efforts in this area is shown in Fig. 2.3.[1]

Problem B: Chemical clusters. One standard example for global optimization algorithms comes from the problem of finding the lowest energy configuration of n molecules, where n is somewhere in the range of a few to a few thousand. This represents a problem of considerable interest to the chemical and solid state physics communities. Such systems, called clusters, provide a link between isolated molecules and bulk matter [Ber93, BBK+96]. The toy version we use for illustration involves n identical particles whose total energy is given by a sum of spherically symmetric pair potentials

$$E(r_1, r_2, \ldots, r_n) = \sum_{i<j} f(\|\vec{r}_i - \vec{r}_j\|), \tag{2.3}$$

where $\vec{r}_i$ is the position vector of the ith particle, $\|\,\|$ is the Euclidean norm in $\mathbb{R}^3$, and f is either the Lennard–Jones potential,

$$f(\|\vec{r}_i - \vec{r}_j\|) = \frac{-A}{\|\vec{r}_i - \vec{r}_j\|^6} + \frac{B}{\|\vec{r}_i - \vec{r}_j\|^{12}}, \tag{2.4}$$

[1] This figure was kindly provided by the Danish firm Ødegaard, whose product, ISIS, has been offering seismic inversion techniques using global optimization to the oil industry since the early 1990s.

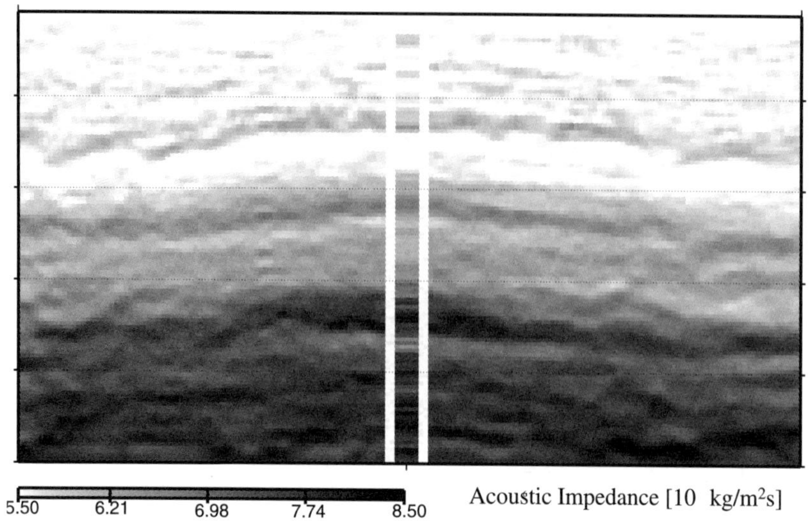

Figure 2.3. *A sample picture of the layers beneath the earth's surface obtained using seismic inversion. The horizontal scale is* 4 *km while the vertical scale is* 500 *m. The gray scale indicates acoustic impedance, which is a product of the sound velocity and the density. The strip in the middle shows the correct answer, i.e., the results from a direct measurement in a well at that position.*

or the Morse potential,

$$f(\|\vec{r_i} - \vec{r_j}\|) = A(1 - \exp(-B\|\vec{r_i} - \vec{r_j}\|)), \tag{2.5}$$

where A and B are positive constants. The problem is to find a set of positions r_i, $i = 1, \ldots, n$, that minimize E. Real versions of the problem involve more complicated pair potentials that depend on orientation as well as distance and often require quantum mechanical formulations [YBS91]. An optimal configuration of about 100 particles is shown in Fig. 2.4.

Problem C: Spin glasses. Our next example serves as a prototype of a large class of problems in physics and computer science [FH91]. The problem involves n Boolean variables $\sigma_i = \pm 1$. Equivalently, the values of the σ_i's can be taken as 0, 1. Here we use the ± 1 convention, as is most commonly done in the spin glass literature. The objective function is

$$E(\sigma_1, \sigma_2, \ldots, \sigma_n) = \sum_{i<j} J_{ij}\sigma_i\sigma_j. \tag{2.6}$$

With given values of the coupling constants J_{ij}, this is known in computer science as the quadratic assignment problem. For spin glasses, the J_{ij} are random variables that are either normally distributed with zero mean and standard deviation J or discrete and taking on the

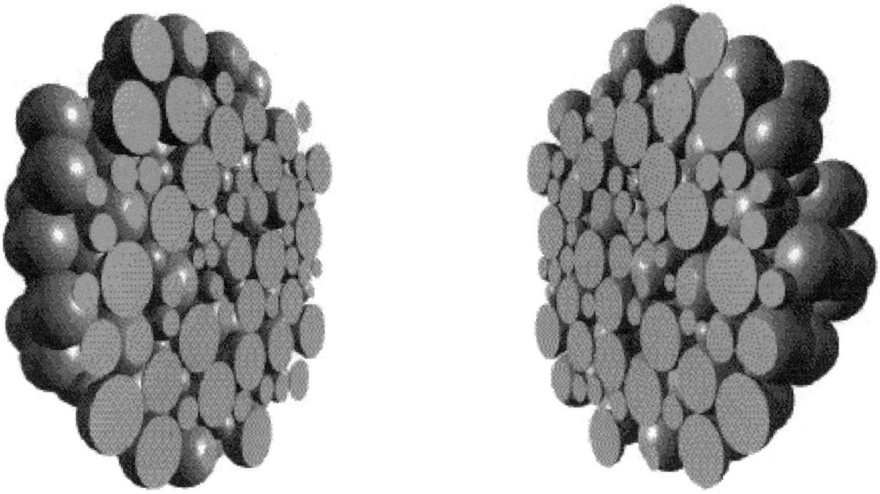

Figure 2.4. *The cross section of a cluster of about* 100 *particles.*

values $+J$ and $-J$. The spin glass problem with the objective function in (2.6) is known as the infinite range spin glass since it allows for interaction between any two spins in the system. Restricted range versions identify each spin index with a specific lattice site in R^3 (or R^2) and take $J_{ij} = 0$ unless site i and j are neighboring sites in the lattice.

A physical realization of a spin glass can be achieved, for instance, by taking a dilute solution of iron in gold. The sign of the actual interaction between spins at the lattice sites occupied by the iron atoms varies sinusoidally with the distance from the site and thus appears random in a dilute solution where relatively few sites are occupied by iron atoms while the other sites are occupied by gold atoms. The random positions of the atoms, and thus the coupling constants J_{ij}, remain fixed in time and are for this reason called quenched variables, in contrast with the magnetic degrees of freedom, which are dynamic variables. The system as a whole is said to possess quenched randomness.

Spin glasses are often used as a paradigm for "frustration"; when a particular spin σ_i is changed from (say) $+1$ to -1, it decreases some terms in the objective function while increasing the values of other terms. In a frustrated system no assignment of signs for the σ_i's exists that minimizes all the terms simultaneously. This system is also important as a caricature of real glasses, which are similarly frustrated in their attempts to find minimal energy configurations on time scales accessible in a laboratory, although they lack quenched randomness.

Problem D: Graph bipartitioning. This example represents one of the standard NP-complete problems [GJ79]. NP-complete is a technical term in computer science that refers to problems that have no known algorithms capable of solving the problem in polynomial time [CLRS2001]. Nevertheless, it appears to be a relatively easy problem for global optimization algorithms in the sense that various greedy algorithms do remarkably well [ZS92].

Consider a graph $G = (V, E)$, where V is a finite set of vertices and $E \subset V \times V$ is a set of edges. G is not a directed graph, so E must be symmetric. This is often denoted by $E = E^t$ and means that $(u, v) \in E$ iff $(v, u) \in E$. The problem is to partition V into two equal-size subsets V_1 and V_2, $|V_1| = |V_2|$, such that the number of edges that connect one subset to the other is minimized. For V_1 and V_2 to partition V, we must have $V_1 \cap V_2 = \emptyset$ and $V_1 \cup V_2 = V$. An edge $e = (u, w) \in E$ connects one subset to the other iff $u \in V_1$ and $w \in V_2$. The objective function to minimize is therefore

$$|\{(u, w) \in E; u \in V_1 \text{ and } w \in V_2\}|. \tag{2.7}$$

In alternative versions of the problem, the constraint that the subsets V_1 and V_2 be equal size is relaxed. To keep the problem interesting, i.e., to prevent the trivial solution $V_1 = V$, $V_2 = \emptyset$, a penalty term proportional to $(|V_1| - |V_2|)^2$ is added to the objective function.[2]

Real problems in this family include circuit partitioning, where the vertices represent circuit elements and the edges represent electrical connections. The motivation is to partition a circuit that is too large to fit onto one chip into two circuits to be placed on two different chips in such a way as to minimize the number of wires connecting the two chips.

We remark that this problem is equivalent to a spin glass with zero magnetization. The equivalence is obtained simply by letting σ be the characteristic function of one of the subsets. In this case, the requirement that $|V_1| = |V_2|$ becomes $\sum_i \sigma_i = 0$.

Problem E: Traveling salesman problem. This example is certainly the best studied and most well known of all global optimization problems. It is also NP-complete. The so-called Euclidean traveling salesman problem can be stated as follows: Given n points $(x_1, y_1), \ldots, (x_n, y_n)$ in the plane, let $D_{i,j}$ represent the Euclidean distance

$$D_{i,j} = \sqrt{(x_i - x_j)^2 + (y_i - y_j)^2} \tag{2.8}$$

between the ith and the jth points. We wish to find a permutation σ of the integers 1 to n that minimizes the sum

$$D_{\sigma(n),\sigma(1)} + \sum_{i=1}^{n-1} D_{\sigma(i),\sigma(i+1)}. \tag{2.9}$$

The n points represent cities along the route of the traveling salesman and the permutation σ specifies the order in which these cities are to be visited. Alternate versions work directly from a distance matrix $D_{i,j}$ without worrying about the intermediary of realization as a set of points in the plane.

Problems of industrial interest include arranging delivery routes of heating oil trucks delivering to homes, pickup routes of milk trucks visiting various dairies, and even positioning of a drill press to the sequence of locations where holes are to be bored in a printed circuit. These real versions often involve side conditions and modified objective functions, such as capacity of the truck and repeated visits back to the depot.

[2]The device of incorporating equality constraints using penalty functions is frequently used to allow constrained problems to be attacked by methods of unconstrained problems [Gre80, Lue84].

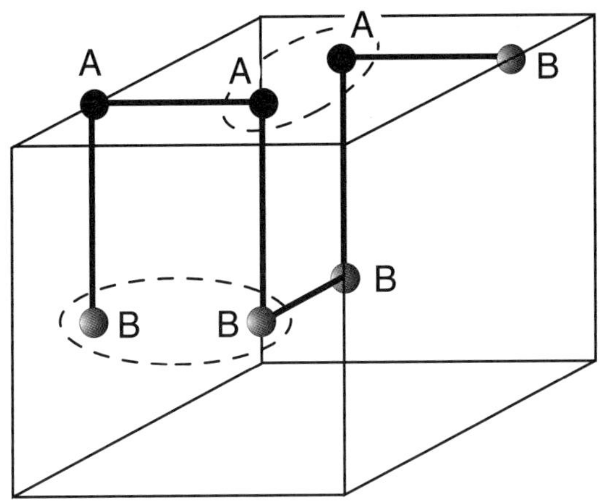

Figure 2.5. *A configuration of a small lattice protein. Like monomers on neighboring sites increase the energy.*

Problem F: Protein folding. This problem is one of the grand challenge problems of contemporary science. In its real form the problem is to start from the linear (primary) structure of the protein specified as a sequence of amino acids and find the three-dimensional configuration of the molecule that minimizes the free energy of the solvated protein in an aqueous solution. Solutions to the problem would be worth millions of dollars to pharmaceutical companies.

The real problem is intractable for many reasons, not the least of which is that it is not known how to accurately calculate values of the objective function.[3] The feeling fueling much of the research effort is that the folding must be rather robust since the real protein is able to find the same optimal configuration time and again. This fact has resulted in researchers referring to a *funneling property* for this problem whereby most search algorithms proceed with overwhelming probability toward the global minimum.

There are many toy versions of the protein folding problem using approximate energies and configurations. The very crude version we consider here uses only two types of amino acid and places each amino acid at a lattice site in $\mathbb{R}^2$ or $\mathbb{R}^3$. Specifically, for a sequence of n amino acids, we seek the (non-self-intersecting) path of length n on the lattice of integer points which minimizes the sum of pair interaction energies between adjacent unbonded amino acids. We illustrate this with a concrete example. Let A and B stand for our two types of amino acids, $n = 7$, and let the sequence to be folded be $BAABBAB$. Using the energy $E(A, A) = 1$, $E(A, B) = E(B, A) = -2$, and $E(B, B) = 2$, the configuration shown in Fig. 2.5 has energy $E_{\text{Total}} = 3$ due to one BB interaction and one AA interaction.

[3]The real objective here would minimize the free energy of a solvated configuration of the protein. While one can certainly write this as a quantum mechanical problem for a given position of all the nuclei in all the atoms in the protein and surrounding water molecules, the calculation of a solution to the resulting equations would be intractable for even one configuration, let alone for the average over positions of solvent molecules that would be required. Accordingly, one must settle for various semiempirical expressions modeling the interactions between the amino acids. These expressions are of course only approximate.

2.2 Move Classes

In Problems A and B, the configuration space over which we optimize is a continuum. By contrast, Problems C, D, and E have discrete configuration spaces with each configuration well separated from any other. Problem F straddles this distinction; the configurations of the real problem live in a continuum while the toy version restricts to a discrete set of configurations. The mathematical tools required for continuous versus discrete variables are usually very different. Nonetheless, for almost every result in analysis, i.e., in the continuous domain, there is a similar and corresponding result in discrete mathematics. As an example, the reader is referred to a discussion of the similarities between difference and differential equations in [Bra66]. In applications of mathematics, there is a rich history of a particular domain of application oscillating back and forth between continuous and discrete models of the same phenomena. This is often discussed in mathematical modeling books where the choice between continuum versus discrete approaches is often based on analytical convenience and tractability. This is also the situation for us, and we will find that there is little difference in our treatment of these two types of problems.

The above description of our problems A–F defines a configuration space over which we optimize and the objective function that is to be optimized. This is enough to define these problems mathematically. A little additional information is needed before these problems can be attacked using simulated annealing—a specification for how the algorithm can move from configuration to configuration as it searches for the optimum. This additional structure is called the *move class* [Whi84] and specifies a rule whereby a configuration can be altered to give another configuration. While almost any rule that will ensure that we can reach all possible configurations will do, the choice of move class can have a dramatic influence on the performance of the algorithm. We have more to say about this in Chapter 10. For now, we will complete the description of problems A–F by specifying a move class to be used with each problem.

Problems A and B. These are our two problems whose space of configurations (the domain of our objective function) are continua. In such problems, this domain takes the form of $\mathbb{R}^k$, i.e., each configuration is specified by the values of k real parameters, $\vec{x} = (x_1, \ldots, x_k)$. For the seismic deconvolution, Problem A, the x's are the variables τ_i and α_i and $k = 2n$. For the chemical clusters, Problem B, the x's are the positions $\vec{r}_i$ of the different particles strung together into a long vector and $k = 3n$.

There is a plethora of reasonable choices of move class for a continuum problem. The simplest move is to change the value of a randomly chosen parameter x_i by $\pm$ a fixed amount. Alternatively, the change could be by a random amount. Fixed-size moves effectively restrict the problem to a rectangular lattice. Random-size moves leave us the freedom of the continuum.

It is somewhat more common to vary all the parameters at once. This is often accomplished by selecting a point in $\mathbb{R}^n$ that is on a sphere of a prescribed small radius centered at the previous state from which the move occurs. Alternatively, we can choose a random point inside the ball of a prescribed radius centered at the old state. If one is not insistent on the Euclidean definition of "ball," this can be achieved by n simultaneous one-variable moves, as described in the previous paragraph—add a uniform random number to each variable. A more common definition of "ball," particularly popular in the studies closely related to physical applications, is an ellipsoid that gives one the freedom to put in the effects of expressing the variables in different physical units. The size of the ellipsoid, rather than being fixed, is often taken to be uniformly distributed about zero. In other words, the move is taken to be a multivariate Gaussian centered at the old state. This has the benefit of occasional large moves with the result that it is less likely to be completely stuck in a local minimum. Additional improvements are claimed to be available from move classes centered at small moves but decreasing more slowly than Gaussian as we move away from the original point. In fact, the distributions used in this capacity decrease so slowly that they have infinite standard deviation. These methods achieved impressive results on several examples [SH87, Ing89, Pen94] and are described more fully in Chapter 10.

Mathematically speaking, the move class defines a graph structure on the set of configurations. The configurations form the vertices of these graphs and edges run between configurations that can be reached from each other by one move. This is the natural way to think about move classes for the discrete problems.

Problem C (continued): Spin glasses. Recall that the configuration space consists of a $+1$ or a -1 for each spin. The most common move class for this problem is to pick a spin at random and change its value.

Problem D (continued): Graph bipartitioning. The standard move involves selecting a random element from V_1 and a random element of V_2 and switching them. This is the natural choice of move class for the problem with the rigid constraint $\|V_1\| = \|V_2\|$. For the penalty function form of the objective, moving single vertices from V_1 to V_2 is sometimes used.

Problem E (continued): Traveling salesman problem. The move class for this problem is often taken to be what are referred to as 2-bond moves. This involves selecting a portion of the tour and reversing how it is connected to the rest. This is illustrated in Fig. 2.6. Specifically, let $\sigma_1, \sigma_2, \ldots, \sigma_n$ be our current state and select at random two integers $1 \leq i < j \leq n$. Then the new tour σ' is given by

$$\sigma_1, \sigma_2, \ldots, \sigma_i, \sigma_{j-1}, \sigma_{j-2}, \ldots, \sigma_{i+1}, \sigma_j, \sigma_{j+1}, \ldots, \sigma_n, \tag{2.10}$$

where all indices must be interpreted modulo n, e.g., if $i = n$, then $i + 1$ must be interpreted as 1.

Problem F (continued): Protein folding. Recall that the state space of the problem is a self-avoiding path of length n. We specify such a path by a sequence a_i, where each a_i takes on one of the values {Up, Down, Left, Right, Forward} coding for how to select the next segment in the walk. The easiest move class to implement is to change a random position to a random value but otherwise keep the rest of the sequence. Note that this gives a perfectly acceptable algorithm for attempted moves which must still be checked for self-avoidance. Another interesting move class is to select two integers $1 \leq i < j \leq n$ and keep intact the portions of the path between 1 and i and between j and n but replace the segment ij by a random path of the same length, i.e., with the same Δx, Δy, Δz, and total path length.

How to cleverly select move classes is an open problem of considerable importance, which is the topic of Chapter 10.

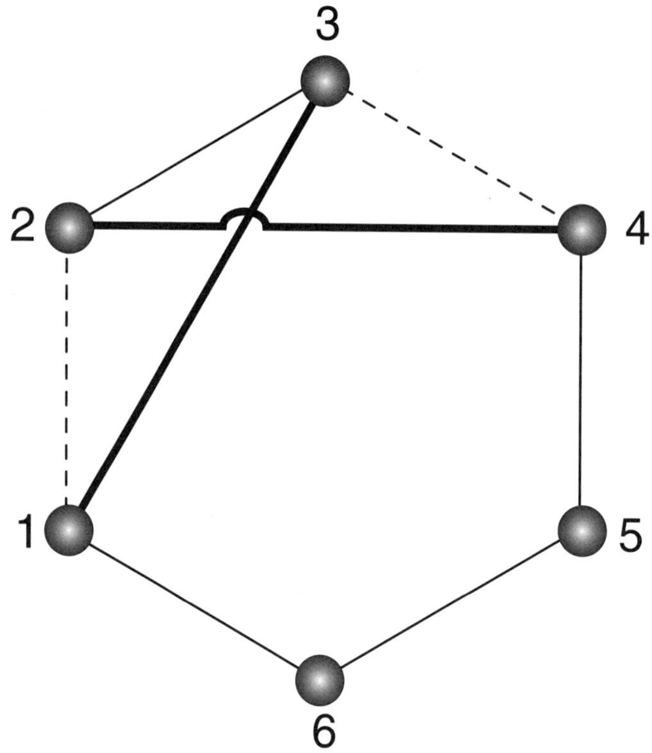

Figure 2.6. *A* 2*-bond move in a small, six-city traveling salesman problem. The fat connections between cities* 1–3 *and* 2–4 *replace the dashed connection between cities* 1–2 *and* 3–4.

Chapter 3
Nomenclature

This brief chapter introduces enough vocabulary and notation to enable readers to bring along their intuition regarding the analogies on which the algorithm is based. The following terms provide an easy reference:

SA	Optimization	Physics	Biology	Symbol
Configuration or state	Feasible solution	Microstate	Genotype	ω or i or ω_i
State space or configuration space	Domain of objective function	State space	Genotype space	$\Omega = \{\omega\}$
Energy	Objective function	Energy	Fitness	E
Move class	"Neighbor of" relation	Perturbed state	Mutation	$\omega' \in \eta(\omega)$
Temperature	Control parameter	Temperature	(not used)[1]	T
Ground state	Global minimum	Ground state	Optimal genotype	ω^*
Landscape	Graph of the objective function	Energy surface	Fitness surface	$\{(\omega, E(\omega))\}$

These terms suffice for the next chapter, which shows how to implement a bare-bones simulated annealing algorithm. Immediately thereafter, however, we shift our point of view to probability distributions on the set of states, and this will require additional nomenclature:

[1] The biological heuristic uses different control parameters.

SA	Optimization	Physics	Biology	Symbol
Ensemble	Many parallel copies of the algorithm	Ensemble	Population	–
Ensemble size	Number of copies	Ensemble size	Population size	N
One iteration	–	One move by the ensemble	One generation	–
State of ensemble	Distribution on the domain	Macrostate	Distribution over genotypes	$\vec{p}$
Random walker	One copy of the algorithm	One copy of the system	Individual	–
Density of states	Number of states as a function of E	Density of states	–	$\rho(E)$

Note that *population* and *ensemble* correspond only in an approximate sense. The individuals in a population are usually interacting whereas the individual copies in an ensemble do not interact.

In addition to the tables above, we include the following terms:

- *basin* is referred to by various names in the literature; however, these do not correspond to the disciplines represented in the table above. A more formal term is the *basin of attraction* of a local minimum. Less formal names for the same concept are *valley, pocket, well, cup* [Haj88], and even *cycle* [Aze92].
- *noise* refers to small errors, perturbations, or fluctuations, depending on the context.

Finally, the terms *random walk* and *Monte Carlo simulation* are treated as synonyms throughout the book.

Chapter 4

Bare-Bones Simulated Annealing

As indicated in Chapter 1, simulated annealing is a heuristic algorithm based on an analogy to physical systems. The next few chapters present enough about the physical description of equilibrium and equilibration to understand how and why the algorithm works; in the present chapter we aim for a rudimentary understanding. This provides an easily accessible presentation to the reader who wants to get started experimenting with the algorithm. Our main purpose, nevertheless, is to motivate later chapters.

Summarizing from Chapter 2, we see that mathematically we need the following three ingredients to define a simulated annealing problem:

- A configuration space Ω that is the domain of our objective function. The elements $\omega \in \Omega$ are also called states of our system.
- The objective function E defined on this domain, $E : \Omega \to \mathbb{R}$. For any state $\omega \in \Omega$, $E(\omega)$ is called the energy of ω.
- A graph structure η defined on this domain that specifies which configurations are to be one move apart; $\eta : \Omega \to 2^{\Omega}$, where 2^{Ω} is the collection of subsets of Ω. For any configuration $\omega \in \Omega$, the states in $\eta(\omega)$ are called the neighbors of ω.

Formally, this can be summarized by saying that a simulated annealing problem is an ordered triple $(\Omega, E(\cdot), \eta(\cdot))$ consisting of a set, a real valued function on the set, and a graph structure on the set.[1]

Simulated annealing works by simulating a random walk on the set of states Ω. In this fashion it searches the state space looking for low-energy states. At each instant during the simulation, we have a current state ω from which we

1. select at random a neighbor $\omega' \in \eta(\omega)$,
2. consider whether to move to ω' and proceed from there or to stay at ω and try again, i.e., whether to accept or reject the move to ω'.

[1] Azencott [Aze92] calls this a Markov kernel.

The second step allows us to introduce a bias in favor of moves that decrease the energy. Let $\Delta E \equiv E(\omega') - E(\omega)$ and define moves as uphill and downhill, respectively, according to $\Delta E > 0$ or $\Delta E \leq 0$. To effect our bias, we always accept downhill moves while only sometimes accepting uphill moves which are needed to avoid getting trapped in local minima. Uphill moves of size ΔE are accepted with probability $x^{\Delta E}$, where $x \in [0, 1]$ is a control parameter. Note that with $x = 1$ all moves are accepted, while with $x = 0$ only downhill moves are accepted. For intermediate values of x the chance of accepting an uphill move decreases as x decreases. Simulated annealing proceeds by a random walk which decreases x from an initial value near one to a final value near zero. Note that this means that we spend progressively less time moving uphill as the algorithm proceeds.

Simulated annealing derives from an early (1953) method for the computer simulation of physical equilibrium at a temperature T. The method is known as the Metropolis algorithm [MRR+53] and serves as the workhorse of physical Monte Carlo simulations to this day. It works exactly as described above except that the control parameter x is held fixed. In more conventional physical notation, the parameter x is expressed in terms of the temperature T by $x = e^{-1/T}$. The core of the Metropolis algorithm is the following procedure for implementing one *Metropolis step*. The procedure is written here as a C or Java-like method `Next_State`, which takes as input the current temperature and state and returns the next state during the random walk.

```
State* Next_State(double T, State* ω)
{

     trial_ω = Select_Neighbor(ω) ;
     ΔE = E(trial_ω) − E(ω) ;
     if (ΔE ≤ 0) then

          Next_State = trial_ω ;

     else
     {

          R =Generate_Random_Number() ;        // uniform on [0, 1]
          if (R <exp(−ΔE/T)) then
               Next_State = trial_ω ;
          else
               Next_State = ω ;
     }
}
```

This procedure calls three other functions: E, which evaluates the energy of a state; `Select_Neighbor`, which selects at random an element from $\eta(\omega)$; and `Generate_Random_Number()`, which generates a pseudo-random number uniform on [0, 1].

Physical Monte Carlo calculations use procedure `Next_State()` to calculate average properties of a physical system at a temperature T by using a procedure like `Metropolis` given below.

```
State* Metropolis(State* start_state, double T, int N_steps)
{

    current_state = start_state ;
    for (int i=1; i≤N_steps; i++)

        current_state = Next_State(T, current_state) ;

    Metropolis = current_state ;

}
```

This algorithm is useful for finding average properties of a physical system that moves in its state space spending some fraction of its time in each state. For a temperature T, the fraction of time spent in any state ω is proportional to $\exp(-E(\omega)/T)$. The usefulness of the Metropolis algorithm derives from the fact that it visits states with exactly these frequencies. In the language of probability, the set of frequencies make up a distribution called the *Boltzmann distribution*, given by

$$\mathcal{P}_T(\omega) = \frac{e^{-E(\omega)/T}}{\sum_{\hat{\omega}\in\Omega} e^{-E(\hat{\omega})/T}}. \tag{4.1}$$

We have much more to say about this distribution in the chapters ahead. For now, we note only that the lower the temperature, the more this distribution favors low-energy states. In the limit $T = 0$, only the lowest-energy state(s) has nonzero probability. Simulated annealing is based on this fact and might thus be called simulated cooling.[2] The bare-bones simulated annealing procedure can then be written as

```
void Bare_Bones_SA()
{

    ω = ω0 ;        // Initialize state
    T = T0 ;        // and temperature.
    for (int N_temperatures=1; N_temperatures≤N_max; N_temperatures++)
    {

        ω = Metropolis(ω, T, N_steps) ;
        T = Next_T(T) ;

    }

}
```

This algorithm uses two constants, N_max and N_steps, as well as an additional procedure called Next_T. Good choices for these constants, for the initialization values of ω_0 and T_0 as well as the Next_T routine, which decrements the temperature, are discussed in Chapter 13.

[2] *Webster's Revised Unabridged Dictionary* defines *anneal* as "to subject to great heat and then to cool slowly."

Part II

Facts

The two chapters in this part present background information from physics and mathematics. Chapter 5 presents the basics of equilibrium statistical mechanics—a theory for the probability that a physical system in thermal equilibrium at a certain temperature occupies any of its possible states. This is the basic paradigm borrowed from physics for simulated annealing. It is also the paradigm for much of information theory. Chapter 6 presents the basics of Markov chain theory, which is the mathematical description of the dynamics of approach to the equilibrium distribution at any temperature. Both chapters present the minimum background needed from these areas for understanding the workings of the simulated annealing algorithm.

Chapter 5

Equilibrium Statistical Mechanics

Statistical mechanics is a large and active field with a rich literature [McQ00]. The following chapters present only the minimal background needed from this subject for understanding the behavior of the simulated annealing algorithm. Such understanding can guide us in tuning the algorithm to the specific needs of particular applications.

At a fixed temperature, simulated annealing mimics the fluctuation behavior of a physical system in thermodynamic equilibrium. This translates in the present context to visiting different states with a probability that decreases exponentially in the energy of the states, that is, according to the Boltzmann distribution. The states of a physical system, sometimes called microstates or microscopic configurations, are a complete specification of the variables of the systems. In physics, the completeness of the description enables one, in principle, to solve the equations of motion by means of classical or quantum mechanics and find the exact time evolution of the variables. In a gas consisting of a large number of molecules a (classical) microstate is a list of the positions and velocities of all the molecules.[1] The key observation underlying statistical mechanics is that, under typical conditions, any dynamical trajectory will pass through, or visit, very many states during the time it takes to perform a macroscopic measurement. Consequently, any measured property will appear to be an average over the properties of many microstates. The fraction of time spent in any microstate can be used to associate it with a probability, thus shifting our point of view from states to probability *distributions* on the set of states.[2] As a result, physical observables become averages over such distributions.

In summary, statistical mechanics replaces the microscopic description of classical or quantum mechanics with a probabilistic dynamics, which generally involves fewer variables. Even more important, statistical mechanics predicts the equilibrium distribution at a temperature T, which typically turns out to be a Boltzmann distribution.[3] Note that in optimization problems we are not forced to use a particular dynamics for the rule whereby

[1] The quantum mechanical description requires specifying all the quantum numbers.

[2] For the classical-mechanical description, this follows from the ergodic hypothesis [Kh49, Huang87].

[3] For a quantum mechanical description this is only approximately true. Taking quantum effects into account leads to two similar distributions—Fermi–Dirac and Bose–Einstein distributions. We encounter the Fermi–Dirac distribution briefly in Chapter 11.

we move from state to state. In Chapter 6, we define and solve a probabilistic dynamics that has the Boltzmann distribution as its prescribed equilibrium solution.

The Boltzmann distribution contains one parameter, the temperature T. At infinite temperature, all states are equiprobable, while, for $T = 0$, only the states of lowest energy have a nonzero probability. There are two important advantages of the Boltzmann distribution for global optimization:

1. We have to cool our system through the full range of temperatures. At each temperature, the Boltzmann distribution is the easiest to simulate in the sense that it can be realized in the largest number of ways (see below).

2. The Boltzmann distribution is uniform on energy slices, i.e., if two states have the same energy, then they have the same probability. This gives us a distribution that is not biased by kinetic factors, i.e., by how rapidly we were able to locate a certain low-energy state.[4]

As the temperature is lowered, the system is cooled and the mean energy decreases. Equilibrating "fully" at each temperature gives the least biased distribution consistent with the mean energy. This chapter is devoted to explaining in what sense this is true and deriving the Boltzmann distribution from the assumption that all states of a system with a fixed total energy are equally likely. The final result can be summed up in the equation

$$\mathcal{P}_T(E) = \rho(E)e^{(-E/T)}/Z. \tag{5.1}$$

In a simulated annealing context, where a population of random walkers moves in configuration space according to suitable probabilistic update rules, this equation gives the fraction of random walkers at energy E after the system has equilibrated at temperature T. Here $\rho(E)$ is the number of states with energy E, and Z is a normalization factor ensuring that the sum of $\mathcal{P}_T(E)$ over all values of E yields one. The remainder of this chapter may be skipped by those readers allergic to mathematical arguments.

5.1 The Number of States That Realize a Distribution

Our derivation proceeds for the case of a discrete state space, i.e., we assume that the microstates can be indexed by the natural numbers $i = 1, \ldots, M$. Rather than carrying the cumbersome notation ω_i, we refer simply to state i. The energies of these states can then be denoted by $E_1, E_2, \ldots, E_M$. A macrostate of our system is a distribution over microstates, with microstate i having probability p_i. As a concrete realization of the distribution at one instant, we employ the device of an ensemble—N copies of our system with each copy in a particular microstate. A state of the ensemble then represents a distribution by having n_i copies of our system in microstate i, with the ratios n_i/N as close as possible to the p_i's. In the limit of an infinite ensemble, $N \to \infty$, this can be achieved with perfect accuracy. We then ask the question of the number of ways that a given distribution of energies can be realized in an ensemble of N copies of the system. We call n_i the occupation number of

[4]This is quite opposite to the philosophy of other algorithms, e.g., genetic algorithms and evolutionary programming. These can give excellent short-term gains with the long-term cost of losing diversity and eventually getting stuck. As we will see, however, eventually every algorithm gets stuck.

state i. The number of ways that a set of occupation numbers n_i can be realized is given by the multinomial coefficient

$$W = \frac{N!}{\prod_{i=1}^{M} n_i!}. \tag{5.2}$$

We pause to examine what this means using a simple illustration. For the purposes of the illustration, we consider the distribution of income among a group of employees at a company. The employees correspond to copies of the system and their salaries correspond to states. Suppose a company maintains $N = 6$ employees, whom we designate with the numerals 1 through 6, and has the salary values $E_1 = 100$, $E_2 = 200$, and $E_3 = 300$. Then the number of ways we can have two people in each income category, i.e., $n_1 = n_2 = n_3 = 2$, is $6!/(2!2!2!) = 90$. The 90 ways to realize this distribution are listed explicitly in Table 5.1 and readers are encouraged to study this list until they are sure exactly what is meant by: "A distribution can be realized a given number of ways."[5]

Often some a priori information is known about the system. For instance, the average value of the energy or other quantities may be known. The basic postulate of statistical mechanics can then be stated as follows:

All microstates that are consistent with our a priori information are equally likely.

Consider a set of occupation numbers n_i. If one way of realizing this set of n_i's is consistent with our macroscopic information, then all W ways must also be consistent. By our postulate, a set of occupation numbers can be realized in W ways, where W is given by (5.2).

EXAMPLE 5.1. Consider a physical system composed of one particle in a box. Divide the box into M cells each having $1/M$ times the volume of the box. For the purposes of this example, we say that the system is in microstate i if the particle's position falls in the ith subdivision. Now consider N identical replicas of the system and compare the number of ways we can have all N particles in the first subdivision, $n_1 = N, n_2 = 0, \ldots, n_M = 0$ versus the number of ways to have N/M particles in each subdivision, $n_i = N/M$ for all $i = 1, \ldots, M$. For $N = 100$ and $M = 10$, (5.2) gives 1 versus $\frac{100!}{(10!)^{10}} \approx 10^{92}$. This example illustrates why we never see all the molecules in a room spontaneously move to one corner of the room.

For large N, the multinomial distribution is extremely sharply peaked and thus states with n_i that maximize W can be realized in overwhelmingly more ways than other states.

The macroscopic information of interest to us is the total energy to be shared among the N copies of our system, or, equivalently, the average energy per copy of the system. We are now ready to ask the central question of equilibrium statistical mechanics:

Which distribution of energies can be realized in the largest number of ways given a fixed total energy shared among a given number of copies of the system?

[5]In a discrete mathematics class, this would be stated as follows: W is the number of ways of placing N distinguishable objects into M distinguishable urns with urn number i containing n_i objects.

Table 5.1. 90 *ways to allocate* 6 *distinguishable objects to* 3 *distinguishable bins with* 2 *objects in each bin.*

$$\binom{6}{2\quad 2\quad 2}$$

12 34 56	23 14 56	35 24 16
12 35 46	23 15 46	35 21 46
12 36 45	23 16 45	35 26 41
12 45 36	23 45 16	35 41 26
12 46 35	23 46 15	35 46 21
12 56 34	23 56 14	35 16 24
13 24 56	24 13 56	36 24 15
13 25 46	24 15 36	36 21 45
13 26 45	24 16 35	36 25 41
13 45 26	24 35 16	36 41 25
13 46 25	24 36 15	36 45 21
13 56 24	24 56 13	36 15 24
14 32 56	25 14 36	45 23 16
14 35 26	25 25 46	45 21 36
14 36 25	25 16 43	45 26 31
14 25 36	25 43 16	45 31 26
14 26 35	25 46 13	45 36 21
14 56 32	25 36 14	45 16 23
15 34 26	26 14 53	46 24 13
15 32 46	26 15 43	46 21 43
15 36 42	26 13 45	46 23 41
15 42 36	26 45 13	46 41 23
15 46 32	26 43 15	46 43 21
15 26 34	26 53 14	46 13 24
16 34 52	34 21 56	56 24 13
16 35 42	34 25 16	56 21 43
16 32 45	34 26 15	56 23 41
16 45 32	34 15 26	56 41 23
16 42 35	34 16 25	56 43 21
16 52 34	34 56 21	56 13 24

The solution of this central question is a straightforward optimization problem that calls for a technique usually covered in the third semester of a standard calculus sequence: Lagrange multipliers. The calculation leads to the Boltzmann distribution.

The logarithm of W is called the entropy of the system.[6] Maximum entropy distributions have proved to be valuable for many applications where they provide a best guess given limited information. The range of applications is extremely broad, and a yearly conference is devoted to new ones.[7] To emphasize the range of examples covered, we mention two examples: the size distributions of fragments in an explosion given the mean fragment size, and the salary distribution in a corporation given the salary levels and the total budget for salaries. The annual conferences, and the corresponding applications, followed Jaynes' seminal exposition showing that this portion of statistical mechanics can be developed using only information theory [Jay83].

5.2 Derivation of the Boltzmann Distribution

We now solve the problem

$$\text{Maximize } W, \text{ subject to} \tag{5.3}$$

$$\sum_{i=1}^{M} n_i = N \quad \text{and} \quad \sum_{i=1}^{M} E_i n_i = E_{\text{Total}}. \tag{5.4}$$

To do the calculation one relies on two tricks. The first is to use what is known as the thermodynamic limit, i.e., to let both N and the n_i's approach infinity, so we may deal instead with the continuous variables $p_i = n_i/N$ representing the fraction of the population in state i. The second trick is to maximize $\ln(W)$ rather than W. Since the natural logarithm is a strictly increasing function, maximizing W is equivalent to maximizing $\ln(W)$. Using the ln function has the benefit of turning the numerous products appearing in W into sums. We begin by changing the objective function to $\ln(W)$ and dividing both sides of the constraint equations by N, giving the equivalent problem

$$\text{Maximize } \ln(W) = \ln(N!) - \sum_{i=1}^{M} \ln(n_i!), \text{ subject to} \tag{5.5}$$

$$\sum_{i=1}^{M} \frac{n_i}{N} = 1 \quad \text{and} \quad \sum_{i=1}^{M} E_i \frac{n_i}{N} = E_{\text{Total}}/N. \tag{5.6}$$

The constraints are then easily expressed in terms of the p_i. To express the objective function $\ln(W)$ in terms of the p_i, we make use of Stirling's approximation for the logarithm of a factorial,

$$\ln(K!) \approx K \ln(K) - K. \tag{5.7}$$

[6] Actually, the entropy is the logarithm of the number of microstates. The sharply peaked nature of the multinomial distribution then gives that the largest W is about equal to the sum of all the W's representing microstates and the entropy is equal to $\ln(W_{\max})$

[7] The 20th annual conference in the series, MaxEnt 2000, was held in Paris.

Using this for $\ln(N!)$ and each of the $\ln(n_i!)$ gives for our objective

$$\ln(W) = N\ln(N) - N - \sum_{i=1}^{M}(n_i \ln(n_i) - n_i). \tag{5.8}$$

The terms $-N$ and $\sum_{i=1}^{M} n_i$ cancel. The first term can be rewritten as

$$N\ln(N) = \sum_{i=1}^{M} n_i \ln(N), \tag{5.9}$$

which allows us to combine it with the other term, giving

$$\ln(W) = -\sum_{i=1}^{M} n_i \ln\left(\frac{n_i}{N}\right) \tag{5.10}$$

$$= -N\sum_{i=1}^{M} p_i \ln(p_i). \tag{5.11}$$

This is the form of W used in information theory [Bri62, Sha49]. Since dividing our objective by the constant N gives an equivalent objective function, we are finally left with the problem

$$\text{Maximize } f(p) = -\sum_{i=1}^{M} p_i \ln(p_i), \text{ subject to} \tag{5.12}$$

$$g(p) = \sum_{i=1}^{M} p_i = 1 \quad \text{and} \quad h(p) = \sum_{i=1}^{M} E_i p_i = E_{\text{Total}}/N, \tag{5.13}$$

where p is the vector of p_i, the variables with respect to which we optimize f. The technique used involves Lagrange multipliers.[8] The first-order necessary condition for the optimum is that there exist two numbers α and β such that

$$\partial f/\partial p_i = \alpha \partial g/\partial p_i + \beta \partial h/\partial p_i \tag{5.14}$$

for each $i = 1, \ldots, M$. Putting in our f, g, and h gives

$$-\ln(p_i) - 1 = \alpha + \beta E_i. \tag{5.15}$$

Solving for the p_i we get

$$p_i = e^{-1-\alpha} e^{-\beta E_i}. \tag{5.16}$$

In the technique of Lagrange multipliers, the constants α and β are to be eliminated using the constraint equations. We can find the expression $e^{1+\alpha}$ directly by substituting the p_i obtained in (5.16) into the normalization condition (5.13). This gives

$$Z = e^{1+\alpha} = \sum_{i=1}^{M} e^{-\beta E_i}. \tag{5.17}$$

[8]This topic is treated in most calculus textbooks.

The sum on the right-hand side of this equation, usually denoted Z, is called the partition function. The constant β can in principle be similarly eliminated using the condition on the mean energy (5.13),

$$\sum_{i=1}^{M} E_i e^{-\beta E_i}/Z = E_{\text{Total}}/N \equiv \langle E \rangle, \tag{5.18}$$

where we introduced the notation $\langle E \rangle$ for the mean energy. Unfortunately, for all but the simplest cases, this serves only to define β as an implicit function of $\langle E \rangle$ which must be solved numerically for β in terms of the mean energy $\langle E \rangle$. The distribution is parametrically specified in terms of β, which we now show can be identified with the reciprocal temperature. It is a well-known property of Lagrange multipliers that their values equal the derivative of the optimal value of the objective function per unit change in the value of the corresponding constraint [Gre80, Lue84], i.e.,

$$\beta = \frac{\partial f_{\max}}{\partial \langle E \rangle}. \tag{5.19}$$

Since the optimal value of $\ln(W)$ is the *entropy* of the system [McQ00], and since the partial derivative of the entropy with respect to the energy is the reciprocal of the temperature [McQ00], we can identify β with $1/T$, where T is the temperature of the system. We are left with the one-parameter family of Boltzmann distributions

$$p_i = e^{-E_i/T}/Z \equiv \mathcal{P}_T(i), \tag{5.20}$$

where we have again adopted the notation $\mathcal{P}_T(i)$ of (4.1) for the Boltzmann distribution at temperature T. Note that $\mathcal{P}_T(i)$ gives equal probabilities to any two states i and j whose energies are equal, $E_i = E_j$. While this gives our distribution over states, it also gives rise to a closely related distribution over possible values of the energy by summing the probability of all states at one energy,

$$P(E) = \textstyle\sum_{\{i|E_i=E\}} \mathcal{P}_T(i) = \rho(E)e^{(-E/T)}/Z \equiv \mathcal{P}_T(E). \tag{5.21}$$

Here we adopted the standard abuse of notation designating this distribution over energies by $\mathcal{P}_T$, which we shall call the Boltzmann distribution over energies. We also introduced the quantity $\rho(E)$, which is the number of states with energy E and which is known as the density of states. Knowledge of $\rho(E)$ for any specific system completely determines the systems' equilibrium properties. For this reason, the density of states will turn out to play a central role in much of the material that follows. The partition function Z can be rewritten in terms of $\rho(E)$ as

$$Z = \sum_{i=1}^{M} e^{-E_i/T} = \sum_{E} \rho(E)e^{-E/T}. \tag{5.22}$$

Writing Z in this form allows us the freedom to rescale $\rho(E)$ by a constant without any effect on $\mathcal{P}_T(i)$. Frequently, $\rho(E)$ is given in a normalized form so its sum or integral over the appropriate range of E values equals one. In this case, $\rho(E) = \mathcal{P}_\infty(E)$, i.e., it coincides with the infinite-temperature Boltzmann distribution.

The Boltzmann distribution occurs in many physical systems and accounts for many everyday phenomena. As a simple example, we mention that the distribution of gravitational potential energy among air molecules predicts an exponential decrease in the density of these molecules with increase in altitude. Thus, combining this with the ideal gas law, there is an exponential decrease in pressure. This decrease is familiar to anyone who has spent any time in mountainous regions. The fact that such decrease is exponential is an empirical fact correctly predicted by the above considerations.

As a second example, the distribution of kinetic energy in an equilibrium collection of gas molecules is also Boltzmann, although here it usually goes by the name of the Maxwell–Boltzmann distribution. Since in these examples the state space is a continuum, the density of states and thus the Boltzmann distribution both take the form of a probability density,

$$\mathcal{P}_T(E)dE = \frac{\rho(E)e^{(-E/T)}dE}{\int \rho(E')e^{(-E'/T)}dE'}, \tag{5.23}$$

where $\mathcal{P}_T(E)dE$ is the volume in the space of microstates (rather than number of microstates as in the discrete case) with energy between E and $E + dE$.

Since the energy of the molecules depends only on their speed, one usually expresses this distribution in terms of the latter quantity. Substituting the density of states [McQ00], i.e., the volume in state space corresponding to states with speed between v and $v + dv$, results in the distribution

$$p(v)dv = 4\pi \left(\frac{m}{2\pi kT}\right)^{3/2} v^2 e^{-mv^2/(2kT)} dv, \tag{5.24}$$

where $p(v)dv$ is the fraction of molecules with speed in this interval, m is the mass of one molecule, and k is Boltzmann's constant.[9]

To summarize, our derivation of the Boltzmann distribution proceeded in three steps. First we considered all possible realizations of the occupation numbers n_i within an ensemble. We assumed that any particular microstate of the ensemble would appear with equal probability, which implies that any particular set of n_i's appears with a probability proportional to the multinomial coefficient of (5.2). Second, using the fact that the multinomial coefficient is very sharply peaked for large system size, we identified the most probable distribution of n_i's as the only relevant one and then proceeded to maximize the multinomial coefficient, with the relevant constraints. Third, we identified the frequency n_i/N with the probability of finding one copy of the system in state i and arrived at the Boltzmann distribution.

The second and third steps may be questionable for small ensembles, such as one encounters in simulated annealing. But what justifies the basic assumption in the first place? Usually assigning equal probabilities to microstates of a closed system (the ensemble itself) is considered as a basic postulate of statistical mechanics. As an alternative, the theory can also be formulated as a maximum entropy principle [Jay83]: the best possible distribution over a set of states is the one that maximizes the entropy, $S = -\sum_i p_i \ln p_i$, given the constraint $\sum_i p_i = 1$, and possibly other constraints, reflecting our a priori knowledge of the system. If we accept the maximum entropy principle, we can apply it directly. Doing this, we

[9]For physical systems, the entropy is $k\ln(W)$ and thus $\beta = \frac{1}{kT}$ with k being Boltzmann's constant. The latter is needed in these systems because E and T have different physical units.

end up maximizing $-\sum_i p_i \ln p_i$ with the constraints $\sum p_i = 1$ and $\sum p_i E_i = E_{\text{Total}}/N$, which again leads to the same conclusions as above without needing to assume that all the n_i are large.

5.3 Averages and Fluctuations

The partition function Z (see (5.17)) contains all information about the mean and variance of the energy distribution in the system. To calculate the average energy, it is convenient to rewrite (5.22) in terms of $\beta = 1/T$ as

$$Z(\beta) = \sum_E \rho(E) e^{-\beta E}. \tag{5.25}$$

Taking the logarithm of both sides and differentiating using the chain rule, one then finds that

$$-\frac{d}{d\beta} \log Z(\beta) = \frac{1}{Z(\beta)} \sum_E \rho(E) e^{-\beta E} E. \tag{5.26}$$

Now using the fact that

$$\langle E \rangle = \sum_E E \mathcal{P}_T(E), \tag{5.27}$$

we find that

$$\langle E \rangle = -\frac{d}{d\beta} \log Z(\beta). \tag{5.28}$$

Similarly, the variance σ^2 of the energy is given by

$$\sigma_E^2 = \langle E^2 \rangle - \langle E \rangle^2 = \frac{d^2}{d\beta^2} \log Z(T). \tag{5.29}$$

An interesting and useful relation between the heat capacity $C(T) = d\langle E \rangle/dT$ and variance σ_E^2 follows directly from this by noting that d/dT equals $-1/(T^2) d/d\beta$. It follows that

$$d\langle E \rangle/dT = -1/(T^2) d\langle E \rangle/d\beta. \tag{5.30}$$

Substituting from (5.28) and making use of (5.29) gives

$$C(T) = \frac{1}{T^2} \frac{d^2}{d\beta^2} \log Z(T) = \frac{\sigma_E^2}{T^2}. \tag{5.31}$$

The heat capacity is the derivative of the mean energy with respect to the temperature. As such, it describes the equilibrium response of the system's energy to small temperature changes. On the other hand, the variance of the energy quantifies the size of the energy fluctuations at a fixed temperature.

We anticipate that (5.31) will be very useful when we discuss how to construct adaptive annealing schedules in Chapter 13. An annealing schedule is just the time dependence $T(t)$ of the temperature in a simulation. The connection to (5.31) comes basically by writing dT/dt as $C(T)^{-1}d\langle E\rangle/dt$.

We now turn to another issue needed later: fluctuations of the mean energy $\langle E\rangle$ due to the finiteness of the ensemble size N. Consider again a statistical mechanical system made up of N parts, each contributing to the total energy in an additive fashion. The average energy as well as its derivative with respect to the temperature will then be of order N, and it then follows from (5.31) that the variance will also be of order N. The relevant measure for the importance of the energy fluctuations is the standard deviation σ_E, divided by the average energy E. As $\sigma_E/\langle E\rangle$ is of order $1/N^{1/2}$, it follows that fluctuations are unimportant when N is large.[10]

When energy fluctuations are small, the probability density $\mathcal{P}_T(E)$ given by (5.21) is very strongly peaked around its maximum at $E = \langle E\rangle$. In this situation a useful Gaussian approximation[11] to $\mathcal{P}_T(E)$ can be obtained by using the variance from (5.31),

$$\mathcal{P}_T(E) \propto \exp\left(-\frac{(E-\langle E\rangle)^2}{2T^2C(T)}\right). \tag{5.32}$$

At least in the limit of large temperatures, $C(T)$ is often temperature independent and the temperature alone determines the size of the energy fluctuations. Conversely, if the Gaussian fluctuations in the energy of any system are known, one may translate these into a "noise temperature," by means of (5.32).

[10] For some physical systems undergoing phase transitions, special values of the temperature exist, the so-called critical temperatures, where the relative size of fluctuations diverges and where our line of reasoning does not apply.

[11] Recall that the Gaussian distribution of a variable x with mean μ and standard deviation σ is $\frac{1}{\sqrt{2\pi}\sigma}\exp(\frac{(x-\mu)^2}{2\sigma^2})$.

Chapter 6

Relaxation Dynamics—Finite Markov Chains

A *dynamical* model describes the time evolution of a system. By contrast, equilibrium models are static and can usually be recovered from dynamical models by setting all time derivatives to zero. For statistical mechanical systems, *relaxation* dynamics refers to a description of how a system approaches equilibrium. This is exactly the part we need to understand for simulated annealing, which works by simulating successive relaxations of a fictitious physical system at progressively lower temperatures.

Standard models of relaxation dynamics for physical systems in thermal contact with a heat bath [Kam92] involve a Markov process to model the time evolution. Formally, the randomness enters because part of the system variables, the so-called bath degrees of freedom, are described only by their statistical properties rather than are treated exactly by quantum or classical mechanics.

In this chapter, we restrict our discussion to models in which time is discrete. This means that our system evolves by a sequence of jumps or transitions between states and that these transitions occur at epochs marked by the ticks of a clock. For global optimization, the ticks of such a clock are measured in number of function evaluations. A Markov process involving a discrete time variable is called a *chain*. Simulated annealing is a Markov chain model. To specify the Markov chain, we need to specify the *transition probabilities*, i.e., the probabilities that the system will be in each of its possible states at the next instant given its current state.

If these jump probabilities to new states are constant in time, the chain is called *time homogeneous*. For finite chains, time homogeneity (together with a couple of other technical conditions discussed below) implies *stationarity*. The latter means that averages eventually approach constant values and that two-time averages (correlations) depend only on the difference of the time arguments. At any fixed temperature, simulated annealing is a stationary Markov chain. When the temperature is decreased, the transition probabilities change and the chain is no longer stationary. We can learn a great deal, however, about the behavior of the algorithm by understanding stationary chains.

An important class of examples for simulated annealing involves discrete state spaces that are large but finite. For example, this is the case in Problems C, D, and E described in Chapter 2. In this case, simulated annealing becomes a finite Markov chain. For simplicity, we restrict our analysis to this case, although many simulated annealing problems (for example, Problems A and B in Chapter 2) involve state spaces that are continua. The distinction is not important for our purposes and finite chains are accessible with less technical machinery.

6.1 Finite Markov Chains

This section is a brief summary of topics that are optional for many courses and that are, alas, often left uncovered.

Our first example is arguably the simplest model of a random process: a system with a finite number of possible states that moves from one state to the next at the ticks of a clock. A chain is defined by specifying its matrix of transition probabilities, M. This matrix must have all nonnegative entries and each of its columns must sum to one.[1]

$$M_{i,j} = \text{prob}(state(t+1) = i \mid state(t) = j\ ). \tag{6.1}$$

The entries in each column summing to one is a mathematical expression of the fact that the system must be in some state at time $t + 1$.

$$\sum_i M_{i,j} = 1, \quad M_{i,j} \geq 0. \tag{6.2}$$

EXAMPLE 6.1. As a very simple example of a Markov chain that may have some intuitive appeal, consider a supermarket lottery in which one of two tokens is received by a customer at each transaction. The supermarket prints 99 copies of token 1 for each copy of token 2 printed and hands them out at random. The state of our shopper is then designated by one of the following:

- s_1, having no tokens,
- s_2, having only token 1,
- s_3, having only token 2,
- s_4, having both tokens 1 and 2.

The transition probability matrix is

$$\begin{bmatrix} 0 & 0 & 0 & 0 \\ .99 & .99 & 0 & 0 \\ .01 & 0 & .01 & 0 \\ 0 & .01 & .99 & 1 \end{bmatrix}.$$

A state such as s_4 in the previous example is called *absorbing* because once the system moves to s_4, it can never leave.

[1]An arbitrary convention is built into this statement and the reader is warned that many books on Markov chains use the opposite convention, which defines the right side of (6.1) to be $M_{j,i}$. This results in the transpose of all our equations, e.g., rows summing to one instead of columns. See, for example, the book by Kemeny and Snell [KS60], where this convention is used.

EXAMPLE 6.2 (Problem F continued). The state space for a traveling salesman problem consists of all tours. Thus for an n city problem, the state space has $(n-1)!/2$ states and the transition probability matrix is $(n-1)!/2$ by $(n-1)!/2$ with the i,jth entry specifying the probability that during the execution of our annealing algorithm the ith tour is followed by the jth tour. Fortunately, this matrix need not be stored in the computer's memory to implement the simulated annealing algorithm but can be generated one entry at a time as the simulation proceeds. Although the transition matrix for this problem is very large even for moderate n, it is also sparse. If we are using 2-bond moves, there are exactly $\binom{n}{2}$ neighbors of any state.

EXAMPLE 6.3 (Metropolis algorithm). The transition probabilities M_{ij} from state j to state i implicit in the Metropolis algorithm are the following:

$$M_{i,j} = \begin{cases} 0 & \text{if state } i \text{ and state } j \text{ are not neighbors,} \\ \frac{1}{\#\eta(j)} & \text{if state } i \text{ and state } j \text{ are neighbors, } E(i) \leq E(j), \\ \frac{\exp(-\Delta E/T)}{\#\eta(j)} & \text{if state } i \text{ and state } j \text{ are neighbors, } E(i) > E(j), \\ 1 - \sum_{k \neq j} M_{k,j} & \text{if } i = j, \end{cases} \tag{6.3}$$

where $\#\eta(j)$ is the number of neighboring states to state j.

Note that the transition probabilities M_{ij} in the matrix combine the effects of the two independent decisions needed during the algorithm; they are the product of the probability that state i is selected as a candidate (0 or $1/\#\eta(j)$) and the probability that the attempted move to state i is accepted ($\exp(-\Delta E/T)$). The diagonal entries represent the probability of moving from state j to state j, which occurs by rejecting the attempted move. Thus these diagonal entries are computed by subtracting the probability of accepting moves to other states from unity. At infinite temperature, these diagonal entries are zero since all attempted moves are accepted. With the temperature equal to zero, all attempted moves out of a local minimum are rejected with probability one. Hence, local minima are absorbing states for a Markov chain defined by a Metropolis algorithm. At any positive temperature, there are no absorbing states.

A *state of a Markov chain* is a probability distribution over the states of the system. We denote any such distribution by a column vector p whose ith entry p_i is the probability that the chain is in state i. Thus the entries must be nonnegative and must sum to one. The dynamics of Markov chains is simply to multiply the current state of the chain by the matrix of transition probabilities to find the state one time unit later.

EXAMPLE 6.4 (Example 6.1 continued). The initial state for our supermarket example is the vector

$$p(0) = \begin{pmatrix} 1 \\ 0 \\ 0 \\ 0 \end{pmatrix}, \tag{6.4}$$

which represents all customers in state s_1. After one step, i.e., one purchase, the state of the chain is

$$p(1) = \begin{pmatrix} 0 \\ .99 \\ .01 \\ 0 \end{pmatrix} = M * p(0). \tag{6.5}$$

Iterating, we see that $p(n) = M^n * p(0)$. There are two distinct ways to think about how a distribution represents a state of the chain. The first is the point of view of the supermarket for whom the ith entry represents the fraction of customers in state i after n transactions. The second interpretation is the point of view of the individual for whom $p_i(n)$ represents the probability of her being in state i after n transactions. In the language of random walks, these interpretations correspond to the state of a single random walker versus the state of an ensemble of many simultaneous random walkers. We have much more to say about this in the next chapter.

For our study of simulated annealing we are concerned only with a special subclass of chains known as *regular* chains. To guarantee that a chain is regular, we need two assumptions. First we require that any state be reachable from any other state by some sequence of transitions. This condition is called *ergodicity*[2] and guarantees sufficient connectivity for our state space. It is a very reasonable condition to place on the move class of any simulated annealing problem. In this context it is sometimes referred to as transitivity of the move class. Note that ergodicity excludes absorbing states and so strictly speaking ergodicity holds for simulated annealing only if the temperature $T > 0$.

The second assumption is needed to eliminate long-term cycling known as periodicity.

EXAMPLE 6.5. Consider the random walk on the points $\{1, 2, 3\}$ illustrated in Figure 6.1. This has the transition probability matrix

$$\begin{bmatrix} 0 & 1/2 & 0 \\ 1 & 0 & 1 \\ 0 & 1/2 & 0 \end{bmatrix}. \tag{6.6}$$

This is a periodic chain; started in any state, the set of states can be partitioned into two classes: those that can be visited only at even times and those that can be visited only at odd times.

To eliminate periodicity in an ergodic chain, it is sufficient that there exists a state i such that the transitions probability $M_{ii} \neq 0$. Thus periodicity cannot happen in simulated annealing problems since any state that is not a local maximum has at any finite temperature a nonzero probability of rejecting an uphill move and hence remaining where it started. This assures us that the chain of a simulated annealing problem is not periodic.

[2]Our definition is the standard one in Markov chains [KS60]. This definition, in fact, is enough to guarantee the related meaning of ergodicity, which requires that time averages equal averages taken with respect to an invariant distribution.

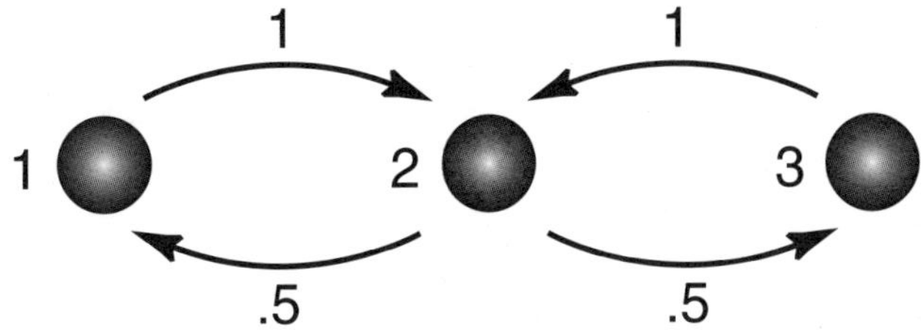

Figure 6.1. *The allowed transitions of a Markov chain on a set of three states. When in state* 2 *the system jumps to the right or to the left with equal probabilities. From states* 3 *and* 1 *the system always jumps to state* 2. *Such a system "blinks" between state* 2 *and states* 1 *and* 3 *and therefore has no equilibrium state.*

Regular chains are chains that are ergodic and not periodic. The following amazing theorem holds for the transition probability matrix M of a regular stationary chain [KS60]:

$$M^n \mapsto M_\infty \quad \text{as } n \mapsto \infty, \tag{6.7}$$

i.e., the powers of the transition probability matrix converge to a limit M_∞. This limit matrix has all of its column vectors equal to the (unique) eigenvector of M with eigenvalue 1. In other words, M_∞ is the matrix $[p_\infty, p_\infty, \dots, p_\infty]$, where

$$M * p_\infty = p_\infty. \tag{6.8}$$

EXAMPLE 6.6. In the land of Simpleweather, each day is either sunny or cloudy or rainy and evolves according to the transition matrix

$$M = \begin{bmatrix} .1 & .3 & .7 \\ .4 & .3 & .2 \\ .5 & .4 & .1 \end{bmatrix}. \tag{6.9}$$

This matrix represents a regular chain. The powers of this matrix tend to the limit

$$M_\infty = \begin{bmatrix} .361842 & .361842 & .361842 \\ .302632 & .302632 & .302632 \\ .335526 & .335526 & .335526 \end{bmatrix}. \tag{6.10}$$

Note that any column of this matrix equals

$$p_\infty = \begin{bmatrix} .361842 \\ .302632 \\ .335526 \end{bmatrix}. \tag{6.11}$$

The physical interpretation of (6.8) says that if we have a system represented by the distribution p_∞, then this distribution does not change, i.e., it remains stationary. Furthermore,

$$M_\infty * p(0) = p_\infty \tag{6.12}$$

for any initial distribution $p(0)$, which says that after a long time the state of the chain approaches p_∞ regardless of the starting distribution. Equation (6.12) means that, started anywhere, a random walker visits all the states with the probability given by this distribution p_∞, which is referred to as the *stationary distribution* of the chain. For a physical relaxation process, such as dipolar molecules of a gas or liquid adjusting their orientation in an electric field, this stationary distribution is the Boltzmann distribution. This is also true of the Metropolis algorithm, i.e., of simulated annealing at any one temperature.

6.2 Reversibility and Stationary Distributions

We are now in a position to verify that the stationary distribution for the random walk implemented in the Metropolis algorithm is the Boltzmann distribution, which is indeed what the algorithm was designed to achieve in the first place. The result follows by noting that the transition probabilities for simulated annealing (see (6.3)) satisfy[3]

$$\frac{M_{ij}}{M_{ji}} = \exp(-\Delta E/T) \tag{6.13}$$

$$= \frac{\mathcal{P}_T(i)}{\mathcal{P}_T(j)}. \tag{6.14}$$

Rearranged, this equation reads

$$M_{ij}\mathcal{P}_T(j) = M_{ji}\mathcal{P}_T(i). \tag{6.15}$$

Consider a very large sample of N random walkers distributed on the set of states according to the Boltzmann distribution, $\mathcal{P}_T(j)$. The number of random walkers in state j is then $N\mathcal{P}_T(j)$. Finally note that M_{ij} is the fraction of the random walkers in state j that move to state i. Thus, (6.15) has the interpretation that the number of random walkers moving from state i to j is exactly equal to the number of random walkers moving in the opposite direction. For this reason, (6.15) is known in the physics literature as the condition of detailed balance. As our interpretation hints, detailed balance is enough to imply that the Boltzmann distribution is the stationary distribution of the Markov chain in simulated annealing.

PROPOSITION. If the transition probability matrix M of a Markov chain satisfies the equation[4]

$$M_{ij}p_j = M_{ji}p_i \tag{6.16}$$

for some probability vector p, then p is the stationary distribution of the chain.

The proposition follows by summing both sides of the given equation over j to obtain

$$\sum_j M_{ij}p_j = \sum_j M_{ji}p_i = p_i \sum_j M_{ji} = p_i. \tag{6.17}$$

[3]We warn the reader of a slight inconsistency in our notation. While $p_i(n)$ represents the probability in state i after n steps, we use $\mathcal{P}_T(i)$ to denote the probability of state i in the Boltzmann distribution at temperature T. The dependence on i is written as a subscript in one case and as functional dependence in the other. We apologize for any confusion that may be caused by this device used to keep the notation manageable.

[4]Note to physicists: We are not using the Einstein summation convention.

In matrix form this equation becomes

$$Mp = p, \tag{6.18}$$

thereby showing that p is the stationary distribution of the chain.

Equivalent and frequent ways to say that a Markov chain satisfies the detailed balance condition (6.16) are that it is *reversible* [KS60] or that it satisfies *microscopic reversibility* [Kam92]. In summary, we have developed the rudiments of the theory of Markov chains and showed that the Markov chain underlying simulated annealing is regular and reversible. At any fixed temperature, the Boltzmann distribution is the stationary distribution of the chain.

There are important similarities and differences between the real dynamics in a physical system and the dynamics represented by the Metropolis algorithm. Real physical dynamics at a fixed temperature must indeed satisfy (6.15), but the M_{ij}'s need not be as in Example 6.3. For each i and j, this leaves open the possibility of an arbitrary common factor for each pair M_{ij} and M_{ji}. Thus, in a move class that begins by selecting a neighbor at random, the move class will satisfy detailed balance provided the probability of accepting the uphill move is $e^{(-\Delta E/T)}$ times as likely as accepting the downhill move. Since the Metropolis algorithm always accepts a downhill move, it is as fast as possible consistent with detailed balance.

6.3 Relaxation to the Stationary Distribution

Recall that the basic dynamics for a stationary Markov chain is given by the equation

$$p(n) = M * p(n-1), \tag{6.19}$$

which may be viewed as a system of first-order homogeneous difference equations with constant coefficients.[5] While the reader is more likely to have seen the basic theorem about such systems in a differential equations class, the theorem for difference equations is almost identical [Bra66] and assures us that the general solution can be written as a linear combination of the eigenvectors of the matrix M, i.e.,

$$p(n) = \sum_{k=1}^{N} a_k \lambda_k^n, \tag{6.20}$$

where λ_k is the kth eigenvalue of the matrix M and the a_k are corresponding eigenvectors.[6] The convergence results stated above regarding powers of M can now be seen in terms

[5]This section assumes slightly more mathematical sophistication on the part of the reader. In particular, we make use of some linear algebra and the theory of a linear system of differential or difference equations. The example at the end of the section illustrates all the concepts on a simple 3-state problem. The main conclusion of the section is that there is a natural time scale for the approach of a Markov chain toward equilibrium. This time scale is denoted by ε and is called the relaxation time of the chain. The reader willing to accept this fact and allergic to mathematics can proceed to the next section.

[6]All the eigenvalues of a reversible chain are real and simple [Kam92]. Without this fact we would have to allow the a_k to be polynomials in n with degree equal to the multiplicity of λ_k as a root of the characteristic polynomial of M.

of the eigenvalues. In this form, the results are widely known as the Perron–Froebenius theorem [Gan59] and state that all the eigenvalues of a regular, aperiodic Markov chain have absolute value less than one except for a single eigenvalue, which equals one. If these eigenvalues are indexed in decreasing order so

$$1 = \lambda_1 > |\lambda_2| \geq |\lambda_3| \geq \cdots, \tag{6.21}$$

then this has the implication that for each $k > 1$, the term $a_k \lambda_k^n$ in (6.20) tends to zero as n tends to ∞. This in turn implies that $p(n)$ tends to a_1, which must therefore be the stationary distribution $p_\infty = a_1$. With these observations, we can restate (6.20) in the following "physical" language. The initial state can be represented as a linear combination of eigenvectors of the matrix M, i.e., vectors v_k such that $Mv_k = \lambda_k v_k$. These eigenvectors are called *modes*, and in this language the above theorem says that all the modes decay to zero geometrically except the stationary distribution, whose contribution to $p(n)$ stays constant.

Our goal is to examine what this implies for relaxation, i.e., for the approach to equilibrium. Since the slowest convergence to zero among the $(\lambda_k)^n$ occurs for the eigenvalue whose absolute value is closest to one, the second largest (in absolute value) eigenvalue λ_2 of the transition matrix M determines the rate of convergence to equilibrium. When $|\lambda_2|$ is much less than one, equilibration is fast. When $|\lambda_2|$ is close to one, equilibration is slow.

There is another, perhaps more common measure of this rate of relaxation based on writing (6.20) in terms of exponentials base e. Let us rewrite

$$|\lambda_2|^n = e^{n \ln(|\lambda_2|)} = e^{-n/\varepsilon}, \tag{6.22}$$

where $\varepsilon = -1/\ln(|\lambda_2|)$ is called the *relaxation time* of the system. Note that slow relaxation corresponds to $|\lambda_2|$ values close to one and thus to very large relaxation times, while fast relaxation corresponds to small values of $|\lambda_2|$ and small relaxation times. ε measures the time scale of the relaxation: each time the number of steps n increases by $\Delta n = \varepsilon$, the distance to equilibrium[7] must decrease by a factor of $1/e$.

EXAMPLE 6.7. To illustrate the ideas in this section, consider the 3-state chain whose transition probability matrix is given by

$$M = \begin{bmatrix} .99 & \frac{1}{2} & 0 \\ .01 & 0 & .1 \\ 0 & \frac{1}{2} & .9 \end{bmatrix}. \tag{6.23}$$

The eigenvalues of this matrix are 1, .9470, and −.0570. The corresponding eigenvectors are

$$\begin{bmatrix} .8929 \\ .0179 \\ .0892 \end{bmatrix}, \begin{bmatrix} 1.0000 \\ -.0860 \\ -.9140 \end{bmatrix}, \begin{bmatrix} -.4775 \\ 1.0000 \\ -.5225 \end{bmatrix}. \tag{6.24}$$

[7] We could alternatively couch this in the language of half-lives by using 2 rather than e as our base. The distance to equilibrium decays with a half life of $\varepsilon/\ln(2)$.

Any initial state, say $\begin{bmatrix} 1 \\ 0 \\ 0 \end{bmatrix}$, can be written as a linear combination of these eigenvectors.

$$\begin{bmatrix} 1 \\ 0 \\ 0 \end{bmatrix} = \begin{bmatrix} .8929 \\ .0179 \\ .0892 \end{bmatrix} + (.1028) \begin{bmatrix} 1.0000 \\ -.0860 \\ -.9140 \end{bmatrix} + (-.0090) \begin{bmatrix} -.4775 \\ 1.0000 \\ -.5225 \end{bmatrix}, \tag{6.25}$$

and it follows that

$$M^n * \begin{bmatrix} 1 \\ 0 \\ 0 \end{bmatrix} = \begin{bmatrix} .8929 \\ .0179 \\ .0892 \end{bmatrix} + (.9470)^n (.1028) \begin{bmatrix} 1.0000 \\ -.0860 \\ -.9140 \end{bmatrix} + (-.0570)^n (-.0090) \begin{bmatrix} -.4775 \\ 1.0000 \\ -.5225 \end{bmatrix}. \tag{6.26}$$

The relaxation time for this example is

$$\varepsilon = -1/\ln(.9470) = 18.36. \tag{6.27}$$

This means that every 18.36 steps we move a factor of $1/e$ closer to equilibrium.

We note an additional general feature illustrated by the eigenvectors in this example. The first eigenvector is a probability distribution—all entries are the same sign and thus by choice of suitable scaling are all nonnegative and add to one. The others have entries with mixed signs that add to zero. The fact that they add to zero allows us to interpret the modes represented by these eigenvectors as a rearrangement of the population.[8]

6.4 Equilibrium Fluctuations

Consider starting a Markov chain from some initial state i_0 at time $t = 0$, i.e.,

$$p_k(0) = \delta_{k,i_0}, \tag{6.28}$$

where the so-called Kronecker delta δ_{k,i_0} equals one if its two indices are equal and zero otherwise. The conditional probability of finding the system in configuration j after t steps, given that it was initially in i_0 is

$$p_j(t) = (M^t p(0))_j. \tag{6.29}$$

This is the jth component of the result of applying the transition matrix M t times to the initial state $p(0)$. Using (6.28), (6.29) can be rewritten as

$$p_j(t) = M^t_{ji_0}, \tag{6.30}$$

where $M^t_{ji_0}$ is the element j, i_0 of the matrix M^t.

[8] The fact that entries of the other transient modes must sum to zero follows in general by summing all the entries in (6.19) starting from different initial distributions. Note that the entries of a_1 sum to 1 since a_1 is a distribution.

According to (6.12), the distribution $p(n)$ approaches the equilibrium distribution p_∞in the limit $t \to \infty$. Thus, all the information regarding the initial state i_0 will eventually be lost. We would expect the deterioration to happen gradually, as the states visited after a few steps should be similar, in some sense, to the initial state. In the rest of this section we try to make these ideas more precise. Technically, we need to introduce the *equilibrium correlation function* of the energy as a measure of the similarity between a state of the system and the state into which it evolves after n steps of the Markov chain. The time decay of the correlation function reflects the way in which the probability distribution relaxes to its equilibrium value.[9]

6.4.1 The Correlation Function

For the present derivation, we assume that the equilibrium distribution $p_\infty(i)$ of the chain is identical with the Boltzmann equilibrium distribution $P_T(i)$ for a given temperature T. To achieve this, the transition matrix M must depend on temperature, although in most of the argument below this dependence is left understood. The dependence is seen explicitly in Example 6.3, which describes the Metropolis algorithm, an algorithm designed to have the Boltzmann distribution as its equilibrium state.

The average energy $\langle E \rangle$ for the state p is

$$\langle E \rangle = \sum_j p_j E(j) \tag{6.31}$$

and will coincide with the thermal average discussed in Chapter 5, if p equals p_∞. The temperature dependence of $\langle E \rangle$ is also left understood.

Assuming that the system is initially in configuration i_0, the *conditional average* of E after t steps is given by

$$\langle E(t) \mid i_0 \rangle = \sum_i E(i) M^t_{ii_0}. \tag{6.32}$$

For $t = 0$, M^t is the identity matrix and $M^t_{ii_0} = \delta_{ii_0}$. Hence, $\langle E, t = 0 \mid i_0 \rangle = E(i_0)$, which is the initial value of the energy. In the opposite limit $t \to \infty$, $\langle E, t \mid i_0 \rangle$ approaches the equilibrium average $\langle E \rangle$ and our knowledge of the initial energy is lost.

Studying equilibrium fluctuations simply means choosing the initial state i_0 according to the equilibrium probability $P_T(i_0)$. In this case, (6.32) takes the form

$$\langle E(t) \rangle = \sum_i \sum_{i_0} E(i) M^t_{ii_0} P_T(i_0). \tag{6.33}$$

Recall that M implicitly depends on a temperature T'. When $T' = T$, $P_T(i_0)$ is the equilibrium distribution for M. Hence $\sum_{i_0} M^t_{ii_0} P_T(i_0) = P_T(i)$ and $\langle E(t) \rangle$ equals the equilibrium average $\langle E \rangle$. If $T' \neq T$ the initial distribution is not stationary, and the left-hand side of (6.33) has an actual time dependence.

[9] Our presentation focuses on the correlation function of the energy. The derivation is easily modified to cover any other quantity of interest.

We define the *energy-energy correlation function* $C_E(t)$ as

$$C_E(t) = \sum_i \sum_{i_0} \langle E(i), t \mid i_0 \rangle E(i_0) \mathcal{P}_T(i_0) - \langle E \rangle^2. \tag{6.34}$$

Initially, the correlation function equals the equilibrium variance of E,

$$C_E(0) = \langle E^2 \rangle - \langle E \rangle^2 = \sigma_E^2, \tag{6.35}$$

while its limit for large t is

$$C_E(\infty) = \langle E \rangle^2 - \langle E \rangle^2 = 0. \tag{6.36}$$

The normalized correlation function defined by

$$c_E(t) = C_E(t)/\sigma_E^2 \tag{6.37}$$

is initially equal to one and eventually decays to zero because the conditional average $\langle E, t \mid i_0 \rangle$ eventually loses its dependence on the initial state i_0. Hence, the magnitude of c_E gauges the extent to which the current state of the system resembles the initial state or, in other words, the extent to which the system remembers its original configuration. In the final state of the decay one can usually assume $c_E(t) \propto e^{-t/\varepsilon}$, where ε is the relaxation time previously introduced.

6.4.2 Linear Response and the Decay of the Correlation Function

The material in this section is somewhat more technical than the rest of the chapter and may be skipped on first reading. The relationships derived here are important for understanding the various adaptive cooling schedules described in Chapter 13.

In an annealing process one decreases the temperature of the system in small jumps, thereby inducing a change in the average energy. Consider a change in the temperature from T to $T - \delta T$. The linear change in the equilibrium average of the energy, or *linear response*, induced by a small change of temperature δT can be written in various ways as

$$d\langle E \rangle = -\delta T \frac{d\langle E \rangle}{dT} = -\delta T \cdot C(T) = -\delta T \frac{\sigma_E^2}{T^2}. \tag{6.38}$$

Here we have used the definition of the heat capacity $C(T)$ and its relation (5.31) to the equilibrium variance of the energy.

Equation (6.38) does not describe how the final value $\langle E \rangle + d\langle E \rangle$ is approached in time after starting from the initial value $\langle E \rangle$. The way in which this happens involves the time dependence of the energy-energy correlation function defined in the previous section.

Consider then a system initially equilibrated at temperature T. At $t = 0$ we instantaneously decrease its temperature to $T - \delta T$. $\mathcal{P}_T$ is no longer the correct equilibrium distribution, and it must then evolve according to a Markov chain, $M = M_{T-\delta T}$, corresponding to the new temperature. The time dependence of the average energy is given by

$$\langle E(t) \rangle_{T \to (T-\delta T)} = \sum_i \sum_{i_0} E(i) M^t_{i i_0} \mathcal{P}_T(i_0), \tag{6.39}$$

which is just (6.33) rewritten with the subscript on the left-hand side to emphasize that the average energy is initially equal to the equilibrium value at T and eventually approaches the equilibrium value at $T - \delta T$.

To make further progress, we need to approximate (again only linearly) the old equilibrium distribution $\mathcal{P}_T(i_0)$ in terms of the new one as

$$\mathcal{P}_T(i_0) = \mathcal{P}_{T-\delta T}(i_0) + \delta T \frac{d\mathcal{P}_{T-\delta T}(i_0)}{dT}. \tag{6.40}$$

We now substitute this into (6.39) to get

$$\langle E(t)\rangle_{T\to(T-\delta T)} = \sum_i \sum_{i_0} E(i) M^t_{ii_0} \mathcal{P}_{T-\delta T}(i_0) + \delta T \sum_i \sum_{i_0} E(i) M^t_{ii_0} \frac{d\mathcal{P}_{T-\delta T}(i_0)}{dT}. \tag{6.41}$$

The first term on the right-hand side is just the mean energy $\langle E\rangle_{T-\delta T}$. To evaluate the second term, we need to evaluate the derivative using the form (5.20) of the Boltzmann distribution and the chain rule as $d/dT = -1/T^2 d/d\beta$. In light of (5.28), this becomes

$$\frac{d\mathcal{P}_{T-\delta T}(i_0)}{d\beta} = \mathcal{P}_{T-\delta T}(i_0)(-E(i_0) + \langle E\rangle_{T-\delta T}). \tag{6.42}$$

Substituting this into (6.41) gives

$$\langle E(t)\rangle_{T\to T-\delta T} = \langle E\rangle_{T-\delta T} + \frac{\delta T}{T^2}\left(\sum_i \sum_{i_0} E(i) M^t_{ii_0} E(i_0) \mathcal{P}_{T-\delta T}(i_0) - \langle E\rangle^2\right). \tag{6.43}$$

As the second term on the right-hand side is nothing but the equilibrium energy-energy correlation function $C_E(t)$ at temperature $T - \delta T$, we finally get

$$\langle E(t)\rangle_{T\to T-\delta T} = \langle E\rangle_{T-\delta T} + \frac{\delta T}{T^2} C_E(t). \tag{6.44}$$

The above equation is a special case of the *fluctuation dissipation theorem*. It tells us that the time-dependent response of the system's energy to a small temperature change is described by the energy-energy correlation function. In particular, the same time scale characterizes the exponential approach of the energy to its equilibrium value as governs the time decay of the correlation function.

As a check, let us see what (6.44) gives at $t = 0$ and $t = \infty$. For $t = 0$ we use (6.35), finding

$$\langle E(0)\rangle_{T\to T-\delta T} = \langle E\rangle_{T-\delta T} + \frac{\delta T \sigma_E^2}{T^2}, \tag{6.45}$$

which agrees with (6.38) since the left-hand side is the equilibrium average of the energy at T. For $t = \infty$, C_E vanishes, and the average on the left-hand side reaches its new equilibrium value at $T - \delta T$, as one expected.

6.5 Standard Examples of the Relaxation Paradigm

This section presents an intuitive treatment of two examples that serve to guide the practitioner in applying statistical mechanics to annealing. The first example concerns equilibration between two microstates, the second between two basins (variously called valleys, pockets, cycles). Each basin consists of several microstates and is qualitatively characterized by the fact that the time scale for escaping the basin is much larger than the time needed to establish an approximate *local equilibrium* within the basin.

6.5.1 Two-State System

Relaxation in a two-state system is a simple and explicitly soluble toy problem, which however provides some useful insight on Markov chains.

EXAMPLE 6.8. Focusing on two states: Consider equilibration between states i and j in a reversible Markov chain. To enable us to focus on what is happening between the two states we restrict moves to exclude all other states. This amounts to looking at the restricted problem whose transition matrix is

$$A = \begin{bmatrix} 1 - M_{ji} & M_{ij} \\ M_{ji} & 1 - M_{ij} \end{bmatrix}, \tag{6.46}$$

which is reversible with the equilibrium distribution

$$\begin{bmatrix} \frac{M_{ij}}{M_{ij}+M_{ji}} \\ \frac{M_{ji}}{M_{ij}+M_{ji}} \end{bmatrix} = \begin{bmatrix} \frac{\mathcal{P}_T(i)}{\mathcal{P}_T(i)+\mathcal{P}_T(j)} \\ \frac{\mathcal{P}_T(j)}{\mathcal{P}_T(i)+\mathcal{P}_T(j)} \end{bmatrix}, \tag{6.47}$$

which is just the equilibrium distribution for the whole chain restricted to states i and j. The larger the coupling between the states, the faster the equilibration. The second-largest eigenvalue is $1 - M_{ij} - M_{ji}$ corresponding to the mode

$$\begin{bmatrix} 1 \\ -1 \end{bmatrix}. \tag{6.48}$$

A number of physically interesting conclusions can be drawn from this example. Relaxation in reversible chains occurs quickly between states with large transition probabilities connecting them. In other words, the relative ratio of the two probabilities reaches the equilibrium value quickly, although the magnitude of both values may be far from what they will be in equilibrium. By this mechanism, the set of states organizes itself into basins in which local stationarity is reached relatively fast. On slower time scales we have equilibration between these basins. Physically, this makes the world as we know it possible with objects (say a piece of iron) lasting long enough en route to their lower basins (rust) to maintain their individuality. In the true equilibrium heat death of the universe, most of the matter is in the form of Fe^{56}. Flow between basins is slow due to barriers. Indeed, there are enormous energetic barriers to the initiation of the nuclear reactions required to transmute the elements into Fe^{56}.

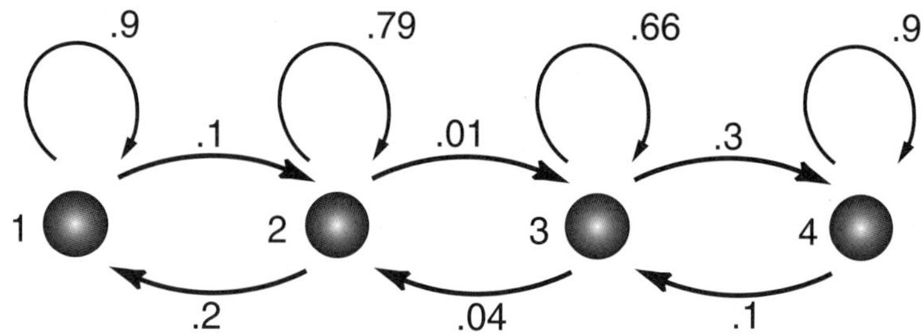

Figure 6.2. *The allowed transitions of a Markov chain on a set of four states. Unlike the previous example, the system has a nonzero probability of remaining in the same state at each time step. Also unlike the previous example, this chain has a stationary distribution.*

EXAMPLE 6.9. This idea of basins is illustrated in the example shown in Fig. 6.2, whose transition probability matrix is given by

$$M = \begin{bmatrix} .9 & .2 & 0 & 0 \\ .1 & .79 & .04 & 0 \\ 0 & .01 & .66 & .1 \\ 0 & 0 & .3 & .9 \end{bmatrix}. \tag{6.49}$$

Since M is nearly block diagonal, two basins B_1 and B_2, consisting of states $\{1, 2\}$ and $\{3, 4\}$, respectively, can immediately be identified. The weak connections between these basins are provided by the transition rates $M_{23} = 0.04$ and $M_{32} = 0.01$, which are much smaller than all other transition rates in the system. The stationary distribution is

$$\begin{bmatrix} \frac{1}{2} \\ \frac{1}{4} \\ \frac{1}{16} \\ \frac{3}{16} \end{bmatrix}, \tag{6.50}$$

but en route to this distribution, the total probability in B_1 is distributed in the $2 : 1$ ratio between states 1 and 2 and the total probability in B_2 in the $1 : 3$ ratio between states 3 and 4. To see this let us examine the chain after 100 steps:

$$M^{100} = \begin{bmatrix} .55 & .55 & .36 & .34 \\ .27 & .27 & .19 & .18 \\ .05 & .05 & .11 & .11 \\ .13 & .13 & .34 & .36 \end{bmatrix}. \tag{6.51}$$

Column i in this matrix represents the distribution after 100 steps when started from state i. Note that columns 1 and 2 are very similar and that the $2:1$ ratio in B_1 and the $1:3$ ratio in B_2 holds to a decent approximation. Note further that this same statement can be made about columns 3 and 4, although these columns are very different from columns 1 and 2. On longer time scales the matrix does reach the correct limit starting from any state. After 1000 steps, each column of M^{1000} equals the vector in (6.50) to six digits of accuracy.

This idea of relaxation among basins is the next paradigm we present. It has reached the status of a folk theorem.

6.5.2 A Folk Theorem—Arrhenius' or Kramers' Law

The idea of basins is predicated on the low temperature behavior of a one-parameter family of reversible Markov chains representing the system at various temperatures. As the temperature is lowered, the uphill probabilities are decreased and some regions cease to communicate very quickly with other regions. These are precisely the regions separated by states whose energies are large compared to the temperature T. Within each of these regions, the Boltzmann distribution relativized[10] among the states is established early and is maintained during the remaining relaxation. There is a folk theorem known in the chemistry literature as Arrhenius' law and in the physics literature as Kramers' law, which says that the rate at which probability leaves a basin is proportional to the probability at the rim of the basin, i.e., to the Arrhenius factor,

$$e^{(-\Delta E_{\text{activation}}/T)}, \tag{6.52}$$

where $\Delta E_{\text{activation}}$ is the difference between the energy at the saddle and the energy at the minimum of the basin. Many variants of this theorem have been proved in a number of specific contexts—that is the reason we refer to it as a folk theorem.

We have already discussed what the theorem means for an ensemble—the population in a basin is Boltzmann distributed with the number of random walkers leaving the basin proportional to the number at the exit energy. For a single random walker the interpretation is a little different. It means that a random walker at a low temperature walking in the basin will have covered the valley many times over before leaving the valley and the time average of her tour represents the relativized Boltzmann distribution in the valley.

This kind of asymptotic analysis is at the heart of many treatments of the subject, including the studies in simulated annealing described in [Aze92]. It allows for a coarser treatment of the dynamics in which the basin is replaced by a single state. In the following example, we present such a treatment. Variants of this example can be found in most chemistry books, usually in association with a diagram like that shown in Fig. 6.3.

EXAMPLE 6.10. The example has three states, which represent two basins and a barrier. If the energies of these states are taken to be 0, 2, and 1 and we adopt the shorthand notation

$$x = e^{-1/T}, \tag{6.53}$$

[10]Renormalized by dividing each probability by sum of the probabilities over this set of states.

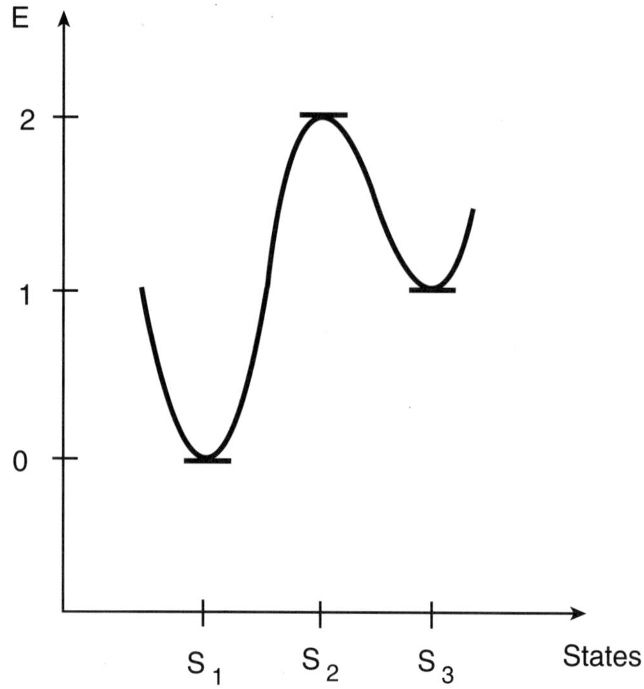

Figure 6.3. *The classical picture illustrating a three-state system, with two minima and a barrier. The abscissa could, for example, represent the reaction coordinate in a chemical system, in which case the ordinate would be the free energy.*

then the transition matrix for our example is

$$M = \begin{bmatrix} 1-x^2 & \frac{1}{2} & 0 \\ x^2 & 0 & x \\ 0 & \frac{1}{2} & 1-x \end{bmatrix}. \tag{6.54}$$

We note that this matrix (with $x = 0$ and with $x = .1$) already provided us with Examples 6.5 and 6.7. The second-largest eigenvalue of the system is

$$\frac{1}{2} - \frac{1}{2}x - \frac{1}{2}x^2 + \frac{1}{2}\sqrt{1 + x^2 - 2x^3 + x^4}, \tag{6.55}$$

whose limiting value as x approaches zero is 1, i.e., whose relaxation time diverges to infinity in the limit of zero temperature. This illustrates the slowing down that is inescapable at low temperatures.

Example 6.10 illustrated energetic barriers between basins. We warn the reader that this is not the only type of barrier in complex landscapes. Equally important are entropic

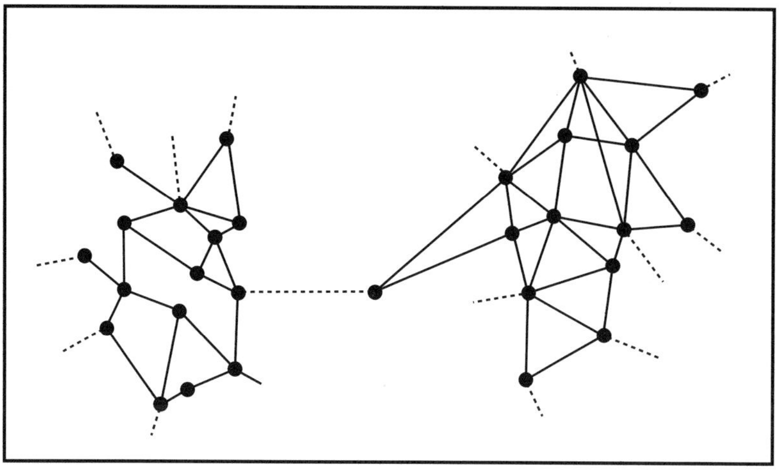

Figure 6.4. *A graph with an entropic barrier. The nodes of the graph fall into two distinct groups, connected by a single edge. A random walker starting in one of the components quickly explores the component itself but has a difficult time finding his way through the bottleneck to the other component. For this reason, the connection can be thought of as an entropic barrier, setting the time scale for the overall relaxation.*

barriers, where it is not the energy that separates two regions but rather the number of possible routes leading from one basin to another: the proverbial bottleneck.

EXAMPLE 6.11. An entropic barrier: Consider diffusion on a planar random graph of the kind illustrated in Fig. 6.4. A diffusing particle, or random walker, is assumed to hop with equal probabilities from any given point to one of its neighbors. If the graph falls into two disconnected components, the random walker will remain confined in the component containing the starting point and the walk is not ergodic. If the structure of the graph is modified with one edge connecting the components, the random walk becomes ergodic. It is intuitively clear that the mean time for traversing this edge can become very large, because the random walker has difficulty finding her way from one side to the other. This constitutes an example of an entropic barrier separating two basins.

6.6 Glassy Systems

The above presentation treated the equilibrium behavior and near equilibrium relaxation of typical thermodynamic systems. Unfortunately, NP-complete optimization problems do not correspond to typical thermodynamic systems. Rather they correspond to glasses—systems whose low temperature relaxation is much slower than the time scale of observation.

The fact that simulated annealing problems represent ergodic chains at any positive temperature was shown in a classic paper by Geman and Geman [GG84]. For typical hard problems their theorem is moot. Real simulated annealing algorithms quickly reach temperatures at which the system is no longer ergodic on the time scales of interest because it takes too long to move from one low-energy region in state space to another. It may be true that there is a nonzero probability of reaching any state from any other state, but once this probability becomes small enough that the chances of seeing such an event during our computer run become negligibly small, the simulation no longer *acts* ergodic.

This is referred to in the physical literature as *broken ergodicity* [Pal82] and is a common feature in what in physical parlance are called glassy systems. These glassy systems have effectively frozen-in disorder, whereby they remain trapped in suboptimal configurations as they are cooled below a certain temperature, called the glass transition temperature. More formally, as a system undergoes a glass transition at temperature T_{glass}, its relaxation time diverges. This can happen in a strict sense only for problems with an infinite number of microstates which in turn means an infinite number of eigenvalues of the transition matrix, some of which might be arbitrarily close to 1. In this case, there is no time scale for which the relaxation can be described as exponential. However, in the laboratory as well as in the computer, infinity merely means large compared to the time scales of observation. If the relaxation time of the system is far greater than the observational time scale, the system is always far removed from thermodynamical equilibrium. The out-of-equilibrium thermalization of glassy systems has several interesting features: often, the behavior of the system changes in a systematic way with the observation time scale. As a rule, at low enough temperatures, simulated annealing problems possess the glass-like behavior that is concomitant to a complex energy landscape. We have more to say on this topic in Chapters 12 and 15.

Part III

Improvements and Conjectures

Myriad interesting versions of simulated annealing can be found in the literature. Most of the variants were introduced to improve the algorithm or to adapt it to a particular application. Some appear to be genuine improvements; some seem to relinquish the connection to the statistical mechanical underpinnings of simulated annealing and can at best be considered ad hoc methods. This part of the book is intended to be an accessible guide to some of the existing options for the novice practitioner. While extensive use is made of the theory developed in Part II, the reader eager to get on with his or her problem can read this part first and refer back to results as needed. In keeping with the title of this book, we restrict our scope to conjectures and demonstrated enhancements of the annealing method itself. Many of these enhancements hinge on the idea of a *natural scale* for a problem and thus are again based on our metatheorem: The more we exploit the structure of a problem, the better.

Basically, the elements of annealing that can be further developed and optimized relative to the simplest bare-bones version already discussed in Chapter 4 are

- the annealing schedule,
- the move class,
- acceptance criterion, and
- the degree and method of parallelization.

The next chapter introduces the idea of ensemble—pooling information from several identical runs ideally performed simultaneously. Chapter 8 deals with a practical problem that naturally arises when using ensembles: how to distribute in the best possible way a fixed amount of computing resources to a set of random walkers. The extreme choices, letting one walker do all the work, and having many walkers do one step each, do not seem intuitively appealing, and indeed it turns out that a compromise can be found leading to a better result.

The remaining chapters of this part of the book deal with more detailed questions like the choice of objective function, annealing schedule, acceptance rule, and move class.

These choices are not really independent of each other, and the performance gain offered by combining them can vary substantially with the type of application. A successful tuning surely entails a certain amount of empirical testing. However, a degree of intuitive understanding of the relationship between performance and the structure of the problem at hand can serve as an indispensable guide to the numerical experiments. Hopefully, the reader has gained such understanding in Part II. Further insight is available in Part IV, where we try to sharpen the reader's intuition in this direction.

Chapter 7

Ensembles

Like all Monte Carlo algorithms, simulated annealing is inherently stochastic, and the way to reproducible and reliable results is to repeat the experiment many times. The statistics gathered from such repeated trials can be used for adapting the algorithm during runtime or in retrospective analysis for preparing future runs. Many of the improvements described in this part of the book rely on information from repeated runs. In keeping with the analogy between simulated annealing and the cooling of a physical system, we refer to such repeated trials as an ensemble. Our use of the term *ensemble* follows standard physical nomenclature to mean either a finite or an infinite number of copies of the system. In biological heuristics, one refers to population for this same meaning. The word ensemble is French for set, and it was first used by Gibbs to denote an infinite number of copies of a system. For theoretical considerations it is usually easier to discuss an infinite ensemble, since frequencies are replaced by probabilities and time evolution becomes deterministic in this limit.[1] However, for the purposes of actual simulations, the ensemble is always finite. The central role played by ensembles in our analysis is due to their simplicity (they have a simple parallel implementation) and to the fact that the collective information extracted from an ensemble is useful for all our ways to improve on the bare-bones annealing algorithm.

In an optimization context, using an ensemble means sharing the search among several random walkers and picking the best result obtained at the end of the run(s). Instead of doing the repeated trials in a sequential fashion, one may choose to perform them in parallel, e.g., on a parallel computer. For this purpose, some authors [BB88, Ban94, Aze92] have advocated partitioning state space and employing a set of random walkers, each restricted to one partition. This seems unnecessary and can even be counterproductive. Such subdivision of tasks is naturally achieved when using an ensemble: At high temperatures the walkers distribute themselves in a random fashion in the landscape. However, when the temperature is lowered, state space falls into dynamically disconnected sets (= valleys), each containing one or more minima of interest. As discussed at the end of Chapter 6, simulated annealing automatically allocates random walkers (members of the ensemble) to these valleys according to the size of the valleys' rims. On the assumption that "deep valleys have large rims,"

[1] Time evolution is deterministic in the sense that given the distribution at some time, the distribution at a later time is uniquely determined.

this should do better than any allocation strategy based on preconceived ideas of splitting up the region.

One reason for considering ensembles arises in connection with sampling problems (Problems A, B, C, and F in Chapter 2), where we are interested in not just the best solution but rather in any good suboptimal solution. This class of problems often has uncertainty or noise in the value of the objective functions (e.g., Bayesian inversion). Ensemble-based methods can provide a distribution of possible answers. The issue of noisy objective functions is dealt with in Chapter 9.

Ensembles are well suited for parallel implementation, which is particularly advantageous if a parallel computer is available. Excluding the possibly substantial parallel speed-ups, the benefits of ensembles can equally well be obtained by doing one search at a time, each search started from a randomly chosen initial configuration. However, an ensemble updated in parallel provides the additional opportunity to use time resolved statistical information adaptively to tune search parameters such as the annealing schedule and the move class. This is done in several of the algorithms described in this part. Incidentally, it should be noted that for adaptive cooling schedules, it does not matter whether the update parallelism is actual or just emulated on a single processor. What matters is that several copies of the system exist in different configurations at any given time during the simulation. Finally, one may choose to go one step further and let the various walkers in the ensemble directly influence each other's search parameters, e.g., by killing bad performers or preferentially breeding good ones [HM97]. With such choices, the search strategy moves to the domain of evolutionary programming and genetic algorithms; the walkers become strongly correlated, and the ensemble idea loses its original meaning.

Chapter 8

The Brick Wall Effect and Optimal Ensemble Size

As discussed in Chapter 4, simulated annealing works by keeping the system in thermal equilibrium[1] while decreasing the temperature. In this way the Boltzmann distribution eventually singles out the ground state of the problem, which is exactly what the optimization procedure is intended for. According to this slightly naive idea, a very gentle and slow cooling—which purportedly keeps the system close to equilibrium—should do better than a rough and quick job. This is indeed the case, up to a point. The phrase *brick wall effect* [HSPS90] describes the frequent situation in which running for a longer time does not lead to noticeable improvements of the solution. In other words, the algorithm is metaphorically stopped by a brick wall.

In Chapter 15 we discuss how the slowing down due to the presence of local energy minima in the energy landscape of the problem generally cannot be avoided. However, trapping effects can be mitigated by a clever allocation of computer time. The method described in this chapter is in fact not restricted to simulated annealing but any Monte Carlo search for the global minimum. Suppose we have at our disposal resources to perform C time steps of a given search algorithm. We could spend all this time to perform a single long search, or—at the other extreme—we could start C different processes, each running a single step, and then choose the best outcome as our putative solution. An intermediate choice confronting us is illustrated by Fig. 8.1, which shows the distribution of best energies seen after C steps and after $C/2$ steps by a random walker. On average, will we be better off choosing once from the first distribution or choosing twice from the second and taking the better of the two choices?

Often it turns out that the clever choice is somewhere between the two extremes. In the context of this resource allocation problem, the brick wall effect can be loosely described as follows:

There is an optimal time to allocate to a single walker.

Resources beyond this amount are best used by starting new, independent searches.

[1]Being in thermal equilibrium means that the time sequence of states visited is (at least approximately) given by the Boltzmann distribution described in Chapter 5.

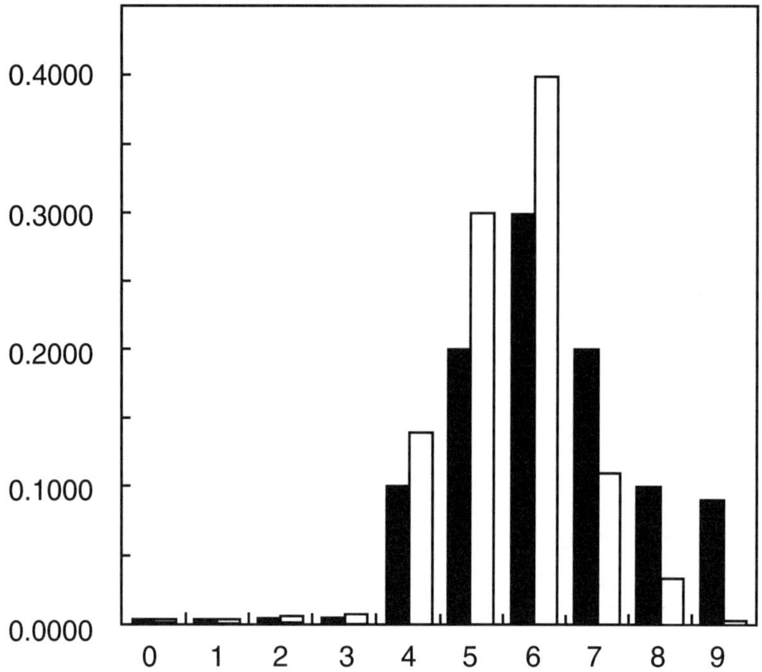

Figure 8.1. *The fraction of random walkers as a function of the lowest energy they have visited after $C/2$ (black) and after C (white) steps. The data shown are discussed in Example* 8.1.

To quantify this statement, we need some of the notation and concepts introduced in Chapters 5 and 6. We consider a random walk of duration t performed by N independent random walkers. Define

$$E_{\text{bsf}}(t) = \text{best-so-far energy} = \min_t E(t) \tag{8.1}$$

as the lowest energy seen by one random walker in time t. Similarly define

$$E_{\text{vbsf}}(t) = \text{very-best-so-far energy} = \min_i E^i_{\text{bsf}}(t) \tag{8.2}$$

as the lowest energy seen by any of the random walkers, where the index i runs over the walkers in the ensemble.

The best-so-far energy is a stochastic process, whose distribution after time t is sampled once by each random walker. Let

$$F_t(E) = \text{Prob}\{E_{\text{bsf}}(t) \leq E\}$$

denote the cumulative distribution of $E_{\text{bsf}}(t)$. F_t is defined for all $E \in \mathbb{R}$ and for fixed t increases monotonically with E from zero to one. If $E_{\text{globalmin}}$ denotes the lowest energy of the system and E_0 denotes the energy of the initial state, then F_t equals zero for all $E < E_{\text{globalmin}}$ and one for $E \geq E_0$. For fixed E, $F_t(E)$ is a nondecreasing function of time.

The distribution of the very-best-so-far energy,

$$G_N(E, t) = \text{Prob}\{E_{\text{vbsf}} \leq E\},$$

can be expressed in terms of F_t by exploiting the fact that the random walkers walk independently and that their respective E_{bsf}'s are identically distributed:

$$\begin{aligned} G_N(E, t) &= 1 - \text{Prob}\{E^i_{\text{bsf}} > E \text{ for all } i\} \\ &= 1 - \prod_{i=1}^{N} \text{Prob}\{E^i_{\text{bsf}} > E\} \\ &= 1 - (1 - F_t(E))^N. \end{aligned} \tag{8.3}$$

With G known, it is a simple matter to quantify the comparison called for in the discussion of Fig. 8.1. The most direct comparison is between the expected or average value $\langle(E_{\text{bsf}}(C))\rangle$ with $N = 1$ and the expected value $\langle(E_{\text{vbsf}}(C/2))\rangle$ with $N = 2$. The general resource allocation problem we wish to solve can be stated as follows:

$$\begin{gathered} \text{Find } t \text{ and } N \text{ that minimize } \langle E_{\text{vbsf}}(t, N)\rangle \\ \text{subject to the constraint } t \cdot N = C. \end{gathered}$$

Let us see how this looks in the context of a simulated annealing problem by considering an example.

EXAMPLE 8.1. Suppose that after binning, the energies in a certain simulated annealing problem take on the values 0 through 9. The measured best-so-far distribution after time t_1 and time $t_2 = 2 \cdot t_1$ are as follows:

E	$p_{\text{bsf}}(t_1)$	$p_{\text{bsf}}(t_2)$
0	0.0010	0.0011
1	0.0020	0.0022
2	0.0030	0.0040
3	0.0040	0.0060
4	0.1000	0.1400
5	0.2000	0.3000
6	0.3000	0.4000
7	0.2000	0.1100
8	0.1000	0.0347
9	0.0900	0.0020

This information is sufficient to compare the average value of E_{vbsf} for $(t, N) = (t_1, 2)$ and $(t, N) = (t_2, 1)$. To carry out the comparison, we have to calculate the expected value of E_{vbsf} for these two choices. For $(t_2, 1)$ the calculation is straightforward. Since there is only one random walker, the E_{vbsf} and the E_{bsf} coincide and thus

$$\langle E_{\text{vbsf}} \rangle = \langle E_{\text{bsf}} \rangle = \sum_{E=1}^{9} E \cdot p_{\text{bsf}}(t_1, E) \tag{8.4}$$

$$= 0 \cdot 0.0011 + 1 \cdot 0.0022 + \cdots + 8 \cdot 0.0347 + 9 \cdot 0.0020 \tag{8.5}$$

$$= 5.554. \tag{8.6}$$

For $(t_1, 2)$ the calculation is a bit more involved. Starting from p_{bsf} we form its cumulative distribution F by taking partial sums

$$F_{t_1}(E) = \sum_{E' \le E} p_{\text{bsf}}(t_1, E'). \tag{8.7}$$

We then calculate the cumulative distribution $G = 1 - (1 - F)^2$ of E_{vbsf} and then take its differences to get $p_{\text{vbsf}} = \Delta G$. The details of this calculation for our example are carried out as follows:

E	$p_{\text{bsf}}(t_1, E)$	$F_t(E)$	$1 - F_t(E)$	$(1 - F_t(E))^2$	$G(E)$	$\Delta G(E)$
0	0.001	0.001	0.999	0.998	0.002	0.002
1	0.002	0.003	0.997	0.994	0.006	0.004
2	0.003	0.006	0.994	0.988	0.012	0.006
3	0.004	0.010	0.990	0.980	0.020	0.008
4	0.100	0.110	0.890	0.792	0.208	0.188
5	0.200	0.310	0.690	0.476	0.524	0.316
6	0.300	0.610	0.390	0.152	0.848	0.324
7	0.200	0.810	0.190	0.036	0.964	0.116
8	0.100	0.910	0.090	0.008	0.992	0.028
9	0.090	1.000	0.000	0.000	1.000	0.008

The $\langle E_{\text{vbsf}} \rangle$ value for $(t_1, 2)$ is 5.425. It is thus better to run two copies for time t_1. Optimizing ensemble size from empirical data typically consists of calculations very similar to the calculation in the present example.

As an alternative objective function for minimization that avoids the integral (or sum) in the expectation value $\langle \rangle$, we can use the median or some other quantile: For $0 < p < 1$, let E_p be the corresponding quantile of G, i.e., $G(E_p, t) = p$. We can choose t in such a way that E_p is minimized, subject to the constraint $t \cdot N = C$. By minimizing the quantile we push the distribution G as far as possible to the left consistent with a given amount of allocated computational resources. In the next chapter we further discuss how the objective function of the annealing algorithm can be chosen.

The problem of ensemble size optimization has been analyzed numerically [JMP88, MV91, HM97] and theoretically [Aze92, HSPS90] for a number of cases. The results from one numerical study are shown in Figure 8.2. The analyses all imply that the optimal time allocated to each random walker, asymptotically for large C, depends very weakly on C, so

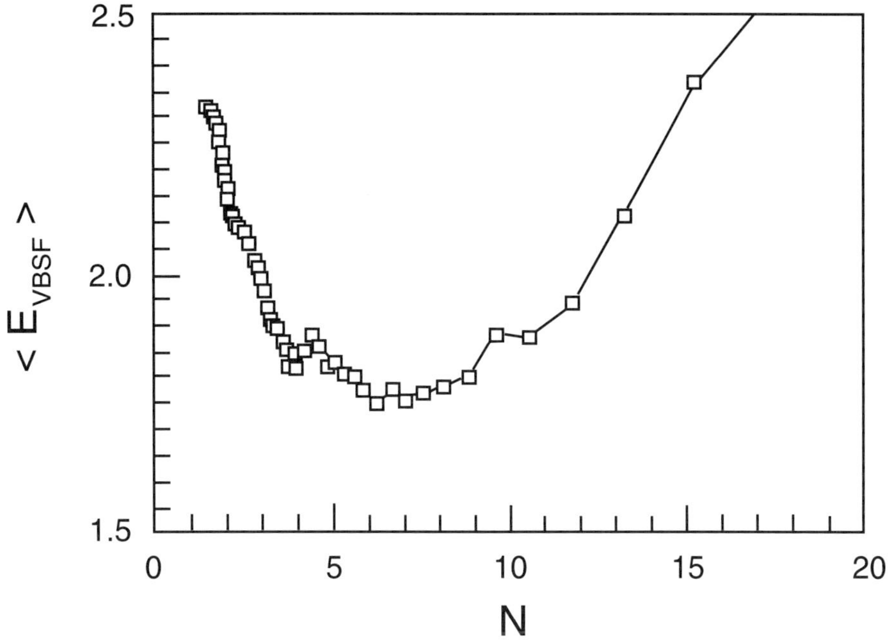

Figure 8.2. *Optimal ensemble size. The expectation value of the lowest energy observed over N parallel optimization processes is shown as a function of N. A clear optimum is seen at around* $N = 6$. *There are* $C = 5.29 \times 10^5$ *function evaluations for each data point. The problem considered is graph bipartitioning on a random graph with connection probability* $p = 0.01$.

that allocation of resources beyond a certain problem dependent amount should be used for independent random walkers. Thus once the optimal time t^* is found for one value of C, the (approximate) solution is easily obtained for any C using the same t^* and $N^* = C/t^*$. While the real problems for which this has been carried out end up with an optimal number N^* of random walkers in the 2–10 range, this number grows linearly with the computer resources and thus we would expect N^* to get sizeable as computing power increases. This shows that for $C \gg t^*$, simulated annealing is nearly *embarrassingly parallel*.[2]

[2]A computation is called embarrassingly parallel if it can be divided into a number of completely independent parts, each of which can be executed by a separate processor.

Chapter 9
The Objective Function

Recall that we are generally interested in very large problems. Our point of view is to do as well as possible with a given amount of computer time. How much computer time one should allocate is ultimately an economic problem that yields easily to a cost-benefit analysis: We need to compare the value of a unit of improvement in the energy with the cost of the expected increase in computational time. As long as it pays more than it costs, we should allocate more time.

How to spend a given amount of time to do "as well as possible" is less clear cut than one might expect. Recalling that the algorithm is stochastic, we realize that our choices of the algorithm and its parameters will determine a distribution of possible values for the best energy found by the algorithm. We might thus conclude that "as well as possible" means that we should minimize the expected value of this best energy. As argued in Chapter 8, where this objective is $\langle E_{\mathrm{vbsf}} \rangle$, beyond a certain level of dedicated computer time, trying to find the best expected value of E_{vbsf} leads naturally to ensembles. This is not, however, the only approach. In fact, for problems in which the objective function is not known with perfect precision, there is a better alternative. This is discussed in the next two sections. Even when the objective is perfectly known, however, there is another approach that modifies (or deforms) the objective as a device for speeding up the convergence of the algorithm. This approach is the topic of the last two sections of this chapter.

9.1 Imperfectly Known Objective

Many important optimization problems come with imperfect information regarding the real objective being minimized. In such problems the values of the objective are accurate only to a certain tolerance. Many of the improvements and conjectures described in Part III distinguish between problems with such noisy objectives and problems in which the objective is known with perfect accuracy. Almost all real problems have some sort of noise in the objective function. We will discuss three sources of such noise.

As our first example of this phenomenon consider the optimal location of emergency facilities in a city.[1] We can minimize the expected response time based on past demand, but our real objective is to meet future demand as well as possible. A related example in which this distinction between historic versus future data plays a prominent role is the tuning of neural networks [HKP91], where a formal distinction is made between the *training energy* (the objective as measured on historical data) and the *generalization energy* (the estimated objective on future data). The goal is to tune the network to optimize performance on future data, and the performance on historic data is only a noisy estimator of this future performance.

Another, perhaps more frequent class of problems with imperfectly known objective comes from problems in which the data gathered suffer from measurement errors. In the seismic deconvolution problem (Problem A), the noise comes from real sound signals in the environment during the measurement of the echoes, from limited accuracy in the microphones used to record the echoes, etc. All these factors limit the quality of match we can expect between the measured and predicted signals, i.e., the accuracy of the energy.

The third class of problems in which the objective is known only to limited accuracy comes from what has been termed "modeling noise" or "modelization noise" [MT95, JP94]. Simple physical models often suffer from this source of error, e.g., ignoring friction in elementary mechanics. In the seismic deconvolution problem (Problem A in Chapter 2) the subsurface does not really consist of perfectly horizontal strata of constant thickness. In the chemical clusters problem (Problem B in Chapter 2), the interparticle interaction energy is not exactly a Lennard–Jones potential. The landscape for folding a protein molecule (Problem F) is much more complicated than can be seen by counting only two classes of amino acids with the interaction energy given in Chapter 2.

9.2 Implications of Noise

In problems without noise, we are usually interested only in the state with the best energy seen and minimizing E_{bsf} makes sense. When dealing with problems with noise we are not concerned with merely finding the very best value of the objective but rather in collecting many alternatives that are nearly equally good. In this case, using a quantile as our objective in selecting an algorithm is perhaps more reasonable. As described in the previous chapter, either objective leads to using ensembles of random walkers with an ensemble size that grows linearly in the amount of computer time.

In problems with noise, one interesting conjecture of how to proceed is due to Mosegaard and Tarantola [MT95]. The conjecture is widely accepted in geophysical inverse modeling, such as seismic deconvolution. The conjecture states that the global minimum is equally likely to be any configuration selected from the Boltzmann distribution at a sufficiently low temperature where the width of the distribution is comparable to the uncertainty in the objective. Mosegaard and Tarantola dubbed this temperature the *noise temperature* and defined it as the temperature at which the mean energy of the system would

[1]This example was used as one of the 1986 problems for the Mathematics Contest in Modeling. For the complete collection, see, for example, Appendix A in [GWF97].

equal the global minimum plus the expected level of fluctuations. (See the discussion concerning fluctuations and Gaussian approximations at the end of Chapter 5.) With this objective in mind, we would tune our algorithm to find a well-equilibrated ensemble at the noise temperature. How large such an ensemble should be again depends on the computer resources.

Consider one consulting firm's approach to doing such calculations for geophysical prospecting customers.[2] After collecting a sample carefully annealed to the noise temperature, they run a clustering algorithm on the population, choose the 10 most populous clusters, construct graphic representations of the subsurface for each of these 10 clusters, and deliver 10 pictures to the customer, each with a probability given by the fraction of random walkers whose energy landed in that cluster.

9.3 Deforming the Energy

One way we might conceive of improving the performance of the algorithm is by deforming the energy function of the problem. Several authors have tried energy distortion methods [Aze92, NA99] which remap the energy monotonically to produce an equivalent objective $\tilde{E}$ in the sense that

$$\tilde{E}(\omega) < \tilde{E}(\omega') \iff E(\omega) < E(\omega'). \tag{9.1}$$

One clear improvement along these lines concerned a problem with large gaps in the range of possible energies [NA99]. In this instance, closing such gaps by remapping the energy improved the performance according to several reasonable criteria, which included faster relaxation times [NA99, NA98].

Azencott [Aze92] studied concave monotonic distortions of the energy and was able to show that, under certain assumptions, such substitutions always increase the rate of convergence to the global minimum.

9.4 Eventually Monotonic Deformations

Another, less restrictive approach is to deform the energy function E to yield a simpler landscape $\tilde{E}_T$ that varies with temperature or some other parameter but eventually matches the original E. In this class are such classic approaches as Stillinger's ant–lion strategy [SW88] and Grest's mean field annealing [GSL86a] and newer so-called homotopy methods [Sun95]. The general idea is to deform the objective to something easily minimized, find the minimum, and then slowly take away the deformation while following the minimum found as it shifts with the removal of the deformation. While this by no means guarantees that we end up at the global minimum, it is usually very fast and gives pretty good results.

Remark. Most primal-dual algorithms work by this principle. In that case the working Lagrangian is just a deformation of the objective function using progressively better values of the dual variables—the Lagrange multipliers.

---

[2]This is the approach taken by the Danish firm Ødegaard in their software package Isis.

We close this chapter by noting that simulated annealing also works in this fashion, albeit at the level of distributions $p = (p_1, \ldots, p_M)$. At any finite temperature, the Boltzmann distribution $\mathcal{P}_T$ minimizes what is known as the free energy,

$$F_T(p) = \langle E \rangle(p) - T S(p), \tag{9.2}$$

where

$$S(p) = -\sum_i p_i \ln(p_i) \tag{9.3}$$

is the entropy of the distribution. F_T is a transformed objective whose optimum at $T = \infty$ is easy to find: it is just the uniform distribution maximizing the entropy. On the other hand, at $T = 0$, F_T coincides with our original objective E, and minimizing it leads to the global minimum. We can therefore view the cooling process as a deformation of the objective function.

Chapter 10

Move Classes and Their Implementations

Recall that in terms of the algorithm, a move class is a procedure that generates a neighboring state to a given state. Problems A–F presented in Chapter 2 included some preliminary discussion of this topic. The issue of selecting possible move classes was discussed very early in the annealing literature [Whi84]. Indeed, clever selection of a move class can make an enormous difference to the rate of convergence to the minimum—perhaps more so than any of the other known improvements. The dramatic successes of some well-designed crossover moves in genetic algorithms come to mind. Note that the use of such crossover moves does not rule out the rest of the simulated annealing algorithm, i.e., Metropolis acceptance with a decreasing temperature.

This chapter is divided into two sections. In the first section we describe some tentative criteria for choosing a move class. In the second section, we discuss alternative move implementations using a given move class.

10.1 What Makes a Move Class Good?

A naive choice of move class is usually easy to make, as discussed in Chapter 2. Nevertheless, assessing the quality of a move class is a problem that, to date, lacks a definitive solution. We note, however, that the choice of a move class has no effect on the stationary distribution of the algorithm at any temperature. As discussed in Chapter 5, the stationary distribution depends only on the number of states at each energy and not on the neighborhood structure, i.e., not on which states are connected to which. The rate of approach to the stationary distribution, however, can be strongly affected by the move class.

Several rules of thumb have been advocated as criteria to construct a good move class; we describe the most attractive of these in the subsections below.

10.1.1 Natural Scales

Many approaches to constructing a schedule are based on the idea of matching the scale of the move to the landscape. Anyone who has ever hiked the Grand Canyon has developed an appreciation for the enormous difference between topographies at different altitudes. Simulated annealing problems are many-dimensional versions of the Grand Canyon. Tailoring the move class to the local energy landscape is in keeping with our often invoked metatheorem: The more we use the structure of the problem, the better. Note that in order to adjust the scale of our moves, it is always possible to arrange for large moves by iterating a given move class, i.e., taking several steps before deciding whether to accept a move. For continuum problems the move size can be easily adjusted up or down.

If, in a given problem, the energy of the configurations varies very little on the scale of one transition, many costly energy evaluations will be required to reach any sizable improvement of the objective function. On the other hand, overly large energy variations in one transition are also undesirable; large moves replace energetic barriers by entropic ones (see Chapters 5 and 15). Rather than fixing the neighborhood size once and for all, many approaches try to capture the fact that the shape of the energy landscape changes with the mean energy, which in turn varies with temperature. The earliest approach due to White [Whi84] advocates choosing the move class at any temperature so the average energy change per step $\langle|\Delta E|\rangle$ equals the temperature. This in fact is similar to the rule of thumb that had been advocated for Monte Carlo techniques before to the invention of simulated annealing—choose the move class so approximately 50% of the attempted moves are accepted [Bin86]. In one empirical study of the traveling salesman problem that reported significant improvement, the authors switched between 2-bond moves and higher rearrangements (3-bond, 4-bond, etc.) while striving to maintain the 50% acceptance rate [LD87]. However, while this approach works well at high to moderate temperatures, it is impossible to implement at low temperatures. This is because once a good configuration has been reached, it is impossible to keep the acceptance rate high enough for these rules of thumb to work; the number of accepted moves becomes very low for any move class. In this low-temperature regime, it pays to employ the more complex schemes described later in this chapter.

10.1.2 Correlation Length and Correlation Time

The configuration space of a system is often described in a picturesque way as an *energy landscape*. The landscape is a graph whose nodes are the configurations of the system. The edges connect neighboring states, and the elementary moves of the dynamics involve crossing one edge.

An interesting way to characterize the landscape geometry was introduced by Weinberg [W90] and further developed by Stadler and collaborators [Sta92, SS92]. Consider an unbiased random walk (i.e., a walk at infinite temperature) performed on the landscape. At each step, the walker visits state ω_t and records the corresponding energy value $E(\omega_t)$. This produces the time series $E(\omega_1), E(\omega_2), \ldots, E(\omega_t), \ldots$. We now assume that this series is stationary and that it fluctuates around the average value μ, with a standard deviation σ.

One can ask how many steps the walker typically performs before losing memory of her initial energy value. Recalling our discussion of the energy-energy autocorrelation function C_E in section 6.4.1, we expect this to be the time scale ξ at which the autocorrelation function of the energy decays to zero.[1] In a number of examples $C_E(\tau)$ is exponential to a very good approximation, i.e.,

$$C_E(\tau) \propto exp(-\tau/\xi), \tag{10.1}$$

and thus fully characterized by ξ. Stadler et al. were able to identify ξ with a purely geometric quantity, the *correlation length* for the energy landscape, which is measured in number of edges and gauges the typical distance at which the energy-energy correlation function decays to zero. This blurs the distinction between correlation length and correlation time for a walk at infinite temperature.

Interestingly, ξ is proportional to the problem size, e.g., the number of vertices in a matching problem [Sta92] or the number of cities in a traveling salesman problem [SS92]. Since the correlation depends on the connectivity of the landscape, and hence on the move class chosen, Stadler et al. suggested that "optimization with general algorithms becomes easier when the correlation of the landscape increases" [SS92]. Thus, by their criterion, move class η_1 is preferable to move class η_2 at a certain temperature T provided

$$\xi_{\eta_1} > \xi_{\eta_2}. \tag{10.2}$$

The expectation behind their hypothesis is that each correlated volume will contain one good minimum only and that the correlation structure will enable the optimization heuristics to quickly find this minimum. Lumping all the states in each correlated volume into just one state leads to a coarse-grained landscape that, ideally, would have no structure. Sorting out which one of the good minima is actually best amounts then to random search, which is a rather poor strategy. Therefore, this part of the search would be kept to a minimum.

Note, however, that to quickly find the best local minimum in a correlated volume with simulated annealing, the energy must quickly relax to its local thermal equilibrium value at a finite temperature T. This would seem to favor the conflicting requirement of a small correlated volume.

10.1.3 Relaxation Time at Finite T

The natural measure of the rate at which we approach the stationary distribution at a certain temperature T is the relaxation time of the Markov chain representing the process. As a consequence of the fluctuation dissipation theorem mentioned in section 6.4, this time corresponds to the autocorrelation time of the walk and captures the scale at which the memory of past configurations fades away.

At low temperatures, many energetically unfavorable moves become difficult to perform. Effectively, the corresponding edges are blocked and the landscape becomes labyrinthine. In this parameter region the relaxation time grows very large and difficult to

[1] For evaluation purposes it is convenient to write the autocorrelation function as an average over a single (in principle infinitely long) trajectory of the system, i.e., as

$$C_E(\tau) = \frac{\langle (E(\omega_t) - \mu)(E(\omega_{t+\tau}) - \mu)\rangle}{\sigma^2}.$$

The equivalence between this expression and (6.34) is a consequence of ergodicity [Kam92].

measure. Its usefulness therefore becomes questionable. Still, at higher temperature the relaxation time can be a useful guide.

Plausibly, a move class with a large correlated volume would have a longer relaxation time. However, for faster relaxation to equilibrium one should choose a move class with a short relaxation time. Hence it seems that the relaxation time criterion is in conflict with the correlation length criterion described in the previous section. Striking the right balance between these two conflicting requirements remains an important subject for empirical studies.

In Chapter 12 we discuss ways in which this relaxation time may be estimated for temperatures T above the glass transition temperature T_g.

10.1.4 Combinatorial Work

Another conjectured measure of the quality of a move class is the number of samples required to select an ensemble of states distributed as $\mathcal{P}_{T+\Delta T}$ by accepting or rejecting samples chosen from the distribution of neighbors of the current states $\eta(\mathcal{P}_T)$ [SHHN88, ZS92]. Here the neighbor of a distribution $\mathcal{P}$ means the distribution obtained by attempting one Metropolis move from a state chosen according to $\mathcal{P}$. Mathematically, this is found by multiplying $\mathcal{P}$ with the infinite temperature transition probability matrix.

The number of samples from $\eta(\mathcal{P})$ that must be examined to select an ensemble of states distributed as a given distribution $\mathcal{R}$ is called the *combinatorial work* of constructing $\mathcal{R}$ from $\mathcal{P}$. This combinatorial work is measured by the Kullback information[2] [Bri62, Kul68] of $\mathcal{R}$ relative to $\eta(\mathcal{P})$. It can be estimated at runtime or retrospectively using the transition matrix method described in Chapter 12.

10.2 More Refined Move Schemes

The techniques described in this section are all helpful in tackling the problem of trapping, i.e., the fact that at low temperatures the system does not thoroughly explore its configuration space but remains close to a local energy minimum. To achieve this, *basin hopping* uses a clever combination of elementary moves, while *fast simulated annealing* uses a distribution of move sizes, allowing occasional long jumps in state space. Finally, *rejectionless Monte Carlo* modifies a given move class by excluding the current state from the set of candidate states. The methods can be combined with any schedule and any move class and they offer considerable speed-ups.

10.2.1 Basin Hopping

An interesting and widely adopted technique is the so-called basin hopping algorithm [WS99], which was applied with excellent results to the chemical cluster problem (Problem B in Chapter 2). The basic idea of basin hopping is to restrict the annealing to the set of local minima. Since this set is much smaller than the full configuration space, one hopefully will obtain a faster convergence. Of course, the improvement comes at a price:

[2]The Kullback information measures the discrepancy of the distribution q_i from the distribution p_i by the quantity $\sum_i p_i \ln(p_i/q_i)$.

the move leading from one minimum to the next becomes more complicated. The move procedure is roughly as follows: Starting from one local minimum, m_1, a state s is chosen (as explained below) and a deterministic minimizer is started from s to reach a second minimum m_2. Finally, this second minimum is accepted or rejected according to the usual Metropolis rule, based on the energy difference $\Delta E = E(m_2) - E(m_1)$. The name "basin hopping" describes the fact that with this composite move class, the system appears to hop from one basin of attraction to another.

How should s be generated? The state s should not be in the same basin of attraction as the original minimum m_1. However, it should not be too far from m_1 either, as one would otherwise risk performing a random search. Jumps of reasonable size adapted to the local topography can be generated adaptively if one is willing to end up at the starting state m_1 for some fraction of the moves. The jump size can be increased every time the local minimizer returns to m_1 and decreased every time the local minimizer ends up at $m_2 \neq m_1$.

The real strength of basin hopping is for continuum problems with differentiable energy surfaces. The greedy part of the algorithm then has access to well-developed tools like conjugate gradient or quasi-Newton, thereby shortening very significantly the time required to explore a basin.

For continuum problems where a very thorough search is desired, an interesting technique for selecting the intermediate state s has produced some excellent results [LD99, WS99]. The technique is called *eigenvector following*, since, starting from m_1, a new nearby point is chosen along the direction of the eigenvector with the smallest eigenvalue of the second derivative of the potential function. Repeating the procedure, i.e., following the eigenvector, one eventually finds a saddle point s when one of the eigenvalues of the second derivative matrix becomes negative. The local minimizer is then started along the direction of negative curvature, eventually leading to the state m_2.

In cases where the problem is discrete or simply too large for the eigenvector following procedure to be practical, any knowledge of the typical length scale characterizing the gross features of the landscape can be utilized to find an appropriate way to jump past the saddle state s.

10.2.2 Fast Annealing

An approach adopted by a significant segment of the simulated annealing community allows (and even relies on) some very large moves. The approach is known as fast simulated annealing [SH87, Ing89] and employs a move class that has its own cooling schedule. The approach selects neighbors from a broad and time-dependent distribution. The direction of the moves is typically uniformly distributed in parameter space, while the change x along a given direction has a Cauchy distribution centered at $x = 0$. This distribution has density

$$P(x) = \frac{1}{\pi} \frac{T_{\text{gen}}}{T_{\text{gen}}^2 + x^2}. \tag{10.3}$$

The shape of $P(x)$ is roughly similar to the Gaussian distribution, with the important difference that its tail dies off more slowly—so slowly, in fact, that the variance becomes infinite. Thus for any positive value of T_{gen}, the algorithm undergoes occasional large-amplitude jumps bringing the system to a nearby basin, which is similar in spirit to basin hopping.

The width parameter T_{gen} of the distribution is decreased according to a cooling schedule, which increases $1/T_{\text{gen}}$ linearly in time and is usually coupled to the cooling schedule for the temperature T used in the acceptance decisions. Fast annealing also modifies the Metropolis acceptance criterion slightly by using a Fermi distribution in place of the Boltzmann distribution. These last two features of fast annealing are discussed further in Chapters 11 and 13.

10.2.3 Rejectionless Monte Carlo

This section describes a way to modify any given move class to counteract the slowing down at low temperatures, where most attempted moves are rejected in conventional schemes.

Rejectionless, or event-driven, methods were pioneered by Bortz et al. [BKL75], who introduced the so-called n-fold way, sampling the Boltzmann distribution over the set of states, in a manner completely equivalent to a Metropolis random walk. The need for rejectionless algorithms has been repeatedly discussed in the literature [Bin79, LB00]. Nevertheless, these methods do not seem to have gained widespread acceptance, although they offer considerable performance improvements over more standard techniques at sufficiently low temperatures.

The basic idea of rejectionless algorithms is as follows: Let α be the current configuration of the system, and let $\beta_1, \beta_2, \ldots, \beta_N$ be a listing of its N neighboring configurations. The usual Metropolis acceptance rule prescribes the probabilities $P(\alpha, \beta_k)$ of jumping from α to β_k. If Δ_k is the energy change corresponding to the jump, this probability is

$$P(\beta_k, \alpha) = \begin{cases} \frac{\exp(-\Delta_k/T)}{N} & \text{for} \quad \Delta_k < 0, \\ \frac{1}{N} & \text{for} \quad \Delta_k \geq 0. \end{cases} \tag{10.4}$$

When α is a local energy minimum all Δ_k's are positive, and the system stays put with probability $1 - \sum_k \frac{\exp(-\Delta_k/T)}{N}$, a quantity which gets close to 1 as T gets small. This regime of low temperatures is rejectionless annealing's domain. The scheme effectively replaces many rejected moves by one move, which represents all the rejections followed by the first success.

The probability of choosing β_k *given that the system jumps* is

$$P(\beta_k, \alpha \mid \text{jump}) = \frac{P(\beta_k, \alpha)}{\sum_j P(\beta_j, \alpha)}. \tag{10.5}$$

Rejectionless annealing simply picks the next state with a probability given by (10.5). The interpretation for the random walk is that this is the state that would be picked after many rejections.

Although this has no direct bearing on optimization per se, one may wish to calculate thermal averages en route, in which case one needs to know the residence time spent at the state α prior to the accepted move. To see how this time is generated, consider the probability Q_α that the system remain unchanged after q attempted moves. This is given by

$$Q_\alpha(q) = \left(1 - \sum_k \frac{\exp(-\Delta_k/T)}{N}\right)^q. \tag{10.6}$$

In the limit of small T (or large N) one obtains

$$Q_\alpha(q) = \exp\left(-\frac{q}{N}\sum_k \exp(-\Delta_k/T)\right). \quad (10.7)$$

Thus, the life time q of state α is exponentially distributed with mean

$$\tau_\alpha = N\left(\sum_k \exp(-\Delta_k/T)\right)^{-1}. \quad (10.8)$$

To update the time after each rejectionless move, one therefore simply increments it by a random number drawn from an exponential distribution with the average τ_α. Our notation needs a slight modification when α is not a local minimum, as a number of exponential factors $\exp(\Delta_k/T)$ in (10.8) must then be replaced by 1's. A variant that has turned out to be particularly efficient for simulating spin glasses [Dall00, Dall01] relies directly on the waiting times associated with each possible spin flip.

Although in principle it is rather simple, the no-rejection approach may require a considerable computational overhead in comparison with standard Metropolis. In a worst-case scenario one must calculate, for each update, N different neighbor energies. However, things are usually more rosy. In the often recurring cases where the energy is a sum of terms depending on different parts of a configuration, a move to a neighbor state only changes one contribution out of N in the overall cost. In these cases, the probabilities of all possible moves can be stored in a list that needs only a few changes at each step. For example, in the spin glass case considered in Chapter 2, such a list would comprise N flip probabilities, one for each spin. After a move is performed that flips spin k, the probabilities that need to be recalculated are those involving the flipping of spin k itself and of those spins with which it interacts. If now the interactions are restricted to spins residing on nearest-neighbor points of a cubic lattice, only six recalculations per update are needed.

Greene and Supovit [GS84] discussed the no-rejection method in a simulated annealing context. They called their method the dynamic weighted selection problem. They found the computational overhead to be of order $\frac{z \log N}{\log z}$, where z is the number of changes in the list induced by a move (i.e., six in our previous example). Since the overhead is independent of temperature, there is a (size- and problem-dependent) temperature at which the standard Monte Carlo becomes less effective than the no-rejection method.

Let's finally remark that since Metropolis without rejections can be combined with any type of schedule, and indeed with almost any kind of probabilistic exploration scheme, it supplements rather than replaces more sophisticated simulated annealing–based schemes.

Chapter 11

Acceptance Rules

Recall that in the Metropolis algorithm a move is made by

1. selecting a neighbor according to the move class, and

2. deciding whether to accept a move to the selected neighbor.

The previous chapter dealt with the first step; this chapter deals with the second. The acceptance rule specified in the Metropolis algorithm is to accept the next state with probability

$$P_{\text{Metropolis}} = \min(1, \exp(-\Delta E/T)) \tag{11.1}$$

and thus depends on the current temperature T. As with all aspects of simulated annealing, there is a significant body of literature exploring variants of this acceptance criterion. We caution the reader that varying the acceptance rule might also vary the resulting equilibrium distribution.

The oldest variant comes not from annealing but from John Holland's classic work [Hol75] leading to genetic algorithms. This family of algorithms is outside the scope of the present book, but we nevertheless mention one simple variant of a population-based acceptance criterion. Select K neighbors (offspring) for each member of a population of size N and make up the population of the next generation by accepting the N lowest energy states of the KN states for breeding at the next iteration. This acceptance rule has the problem that the population loses diversity rather quickly, and it has to be modified to guarantee the possibility of some large moves along the lines of the discussions in the second half of Chapter 10.

The first true variant of the Metropolis acceptance criterion probably dates back to Szu and Hartley's fast annealing (see Chapter 10), which used the acceptance probability

$$P_{\text{FA}} = \frac{1}{1 + \exp(\Delta E/T)}, \tag{11.2}$$

which, like Metropolis acceptance, has the Boltzmann distribution as its stationary distribution at a fixed temperature T. The fast annealing acceptance probability is always less than the Metropolis acceptance probability with the difference becoming negligible for large $\Delta E/T$.

11.1 Tsallis Acceptance Probabilities

Inspired in part by the success of fast annealing, some authors [TS96, Pen94, FH00b, FH00a] introduced a family of rules that occasionally accept large jumps in energy. The family is parametrized by a parameter q, and the acceptance probability is given by

$$P_{\text{Tsal}} = \begin{cases} 1 & \text{if } \Delta E \leq 0, \\ \left(1 - (1-q)\dfrac{\Delta E}{T}\right)^{\frac{1}{1-q}} & \text{if } \Delta E > 0 \text{ and } (1-q)\frac{\Delta E}{T} \leq 1, \\ 0 & \text{if } \Delta E > 0 \text{ and } (1-q)\frac{\Delta E}{T} > 1. \end{cases} \tag{11.3}$$

The stationary distributions generated by P_{Tsal} for $q \neq 1$ are asymptotically equivalent to a family of distributions called Lévy-stable distributions [Fel66b], which have infinite variance and of which the Cauchy distribution (see section 10.2.2) is a special case. Applying L'Hôpital's rule to (11.3), we see that in the limit $q \to 1$ the Metropolis acceptance rule is recovered.

Penna [Pen94] studied the effects of this acceptance rule for the traveling salesman problem. The quality of the results is roughly independent of q, but the speed at which they are reached improves with decreasing q. Their finding led to the theorem discussed in section 11.3.

11.2 Threshold Accepting

A natural modification of the Metropolis acceptance criterion avoids the need to evaluate an exponential function by replacing (11.1) with the approximate form

$$P_{\text{Threshold}} = \begin{cases} 1 & \text{if } \Delta E \leq T, \\ 0 & \text{if } \Delta E > T. \end{cases} \tag{11.4}$$

This approximate form keeps the total probability of accepting an uphill move equal to T

$$\int_0^\infty P(\Delta E) d(\Delta E) = T \tag{11.5}$$

as in Metropolis accepting. Simulated annealing with this acceptance rule is called threshold annealing [DS90, MF90]. The method was originally aimed at providing a poor man's version of simulated annealing by saving the time to evaluate the exponential. This makes the theorem presented in the next section all the more surprising.

11.3 Optimality of Threshold Accepting

Recently Franz and Hoffmann [FH00b] introduced a modified Tsallis acceptance probability

$$\tilde{P}_{\text{Tsal}} = \begin{cases} 1 & \text{if } \Delta E \leq 0, \\ \left(1 - \dfrac{1-q}{2-q}\dfrac{\Delta E}{T}\right)^{\frac{1}{1-q}} & \text{if } \Delta E > 0 \text{ and } \frac{1-q}{2-q}\frac{\Delta E}{T} \leq 1, \\ 0 & \text{if } \Delta E > 0 \text{ and } \frac{1-q}{2-q}\frac{\Delta E}{T} > 1. \end{cases} \tag{11.6}$$

For fixed q, this is equivalent to (11.3) with the rescaled temperature parameter $T' = T(2-q)$. Their modification still reduces to the Metropolis acceptance probability in the limit $q \to 1$ but has in addition property (11.5) and reduces to threshold accepting in the limit $q \to -\infty$. They looked at two toy problems similar to Example 6.10 for which they were able to obtain optimal schedules analytically [FH00a]. Their findings were that the optimal schedules performed best in the limit $q = -\infty$, i.e., for threshold accepting.

As a follow-up study, Franz et al. [FHS01] formulated some general properties that reasonable acceptance criteria should satisfy. They assume that the acceptance probability depends on a parameter T. For $T = \infty$ all moves are accepted. For any fixed T, all reasonable acceptance rules have (by definition of reasonable) the following properties:

(A1) The acceptance probability depends only on the energy difference ΔE.

(A2) For negative energy differences $\Delta E < 0$, the acceptance probability is 1, i.e., downward moves in energy are always accepted.

(A3) For positive energy differences $\Delta E > 0$, the acceptance probability is monotone decreasing, i.e., it is more likely to accept small steps upward in energy than large steps.

(A4) For $\Delta E \to \infty$ the acceptance probability goes to zero, i.e., huge steps upward in energy are very seldom accepted.

Within this framework, the problem of choosing optimal acceptance rules can be formulated. While Franz et al. were not able to solve the problem of finding the optimal rule and the optimal schedule $T(t)$ for the rule, they were able to prove that the optimal rule uses threshold acceptance for each time step. This theorem is the first general statement that has been proved about optimal implementations of finite-length stochastic search algorithms. The proof follows from the fundamental theorem of linear programming, which ensures that the optimal value of a linear function on a set defined by linear inequalities will be taken on at a vertex of the set. The proof holds for a finite state space and for any objective that depends linearly on the final state vector, e.g., the final mean energy or the final probability of being in the ground state.

Knowledge that the optimal performance of a simulated annealing algorithm can be achieved using threshold acceptance is interesting but of limited use without knowing the optimal sequence of thresholds. In particular, it may still be better to use acceptance rules for which a good cooling schedule is known rather than using threshold accepting with a poor schedule.

Chapter 12

Thermodynamic Portraits

This chapter describes how to collect statistical information from an ensemble of independently updated systems. The title of the chapter was coined by Andresen et al. in [AHM+88], where the estimation of parameters characterizing a simulated annealing problem is discussed. The present treatment is more extensive in that it includes additional parameters and reviews a variety of methods for their estimation.

This chapter is divided into sections relating to the three types of information one can gather: equilibrium information, dynamic information, and time-resolved information. All these methods may be viewed as ways to extract natural scales from the problem.

12.1 Equilibrium Information

As shown in Chapter 5, complete equilibrium information about a system is known once we know the density of states, $\rho(E)$. From ρ we can calculate the Boltzmann distribution $\mathcal{P}_T$ at any temperature using

$$\mathcal{P}_T(E) = \rho(E)e^{(-E/T)}/Z \tag{12.1}$$

(see (5.21)). Averaging any property with the Boltzmann distribution tells us that property for the equilibrium ensemble. For example, to find the mean and the standard deviation of the energy at any temperature we use

$$\langle E \rangle_T = \sum_E E\mathcal{P}_T(E) \tag{12.2}$$

and

$$\sigma_E(T) = \sum_E E^2\mathcal{P}_T(E) - \langle E \rangle_T^2. \tag{12.3}$$

For equilibrium systems these two functions of T are related by

$$C(T) = \frac{d\langle E \rangle_T}{dT} = \frac{\sigma_E^2}{T^2}, \tag{12.4}$$

as derived in section 5.3 (see (5.31)).

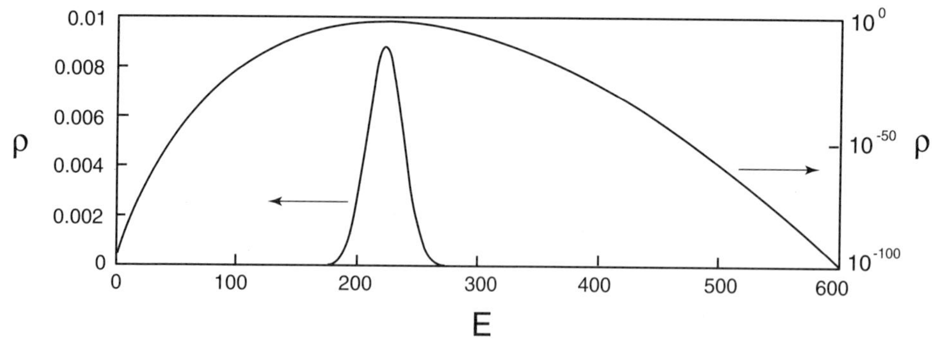

Figure 12.1. *A density of states $\rho(E)$ that varies over* 90 *orders of magnitude. The data shown are for the bipartitioning of a random graph (Problem* D *in Chapter* 2*) with $N = 300$ vertices and edge probability $p = 0.01$. The graph actually shows the binomial distribution $Bi(p, N^2/4)$, which is a close approximation to the actual density of states* [Ped90]. *The lower curve shows $\rho(E)$ in a linear plot, while the upper curve shows a semilog version.*

As discussed in Chapter 13, these two functions $\langle E \rangle$ and σ_E are precisely what is needed for adaptive schedules exploiting the natural energy scale of the system. Both $\langle E \rangle_T$ and $\sigma_E(T)$ can be evaluated in a few different ways. The best way is to use the methods described below to estimate the equilibrium quantities via the density of states, a quantity that is interesting in its own right. Less demanding alternatives evaluate the instantaneous values for the ensemble at the end of a sequence of steps at one temperature or even just use the mean or standard deviation of a time series of energy values for a single random walker.

At first glance, it would seem that one can easily estimate $\rho(E)$ by performing a long enough random walk at any temperature T. The relative frequencies with which different energies are visited is then approximately $\mathcal{P}_T(E) \propto \rho(E)e^{-E/T}$ and thus $\rho(E)$ can be found by multiplying by $e^{E/T}$. This is in general not quite as straightforward as it seems, since $\rho(E)$ typically ranges over many orders of magnitude. Fig. 12.1 shows a typical density of states for an NP-hard problem of moderate size in both a regular and a semilog plot. Note that the values of $\rho(E)$ span about 90 orders of magnitude, i.e., there are about 10^{90} times as many states near $\langle E \rangle_{T=\infty}$ as there are near E_{minimum}. These many orders of magnitude force the runtime estimates of $\rho(E)$ to be very poor for values of E more than a few standard deviations away from the mean value $\langle E \rangle$ at the current temperature. Another problem is that we are not interested in performing long runs at fixed temperatures but rather are interested in cooling the system. This makes it imperative to adopt techniques that can collect information about $\rho(E)$ while sampling at different temperatures during the cooling.

Two commonly used methods exist for estimating $\rho(E)$. The first, known as the *histogram* or *reweighting method*, is the more popular of the two. On the other hand, it does not give us any dynamic information, i.e., it does not give any information about the rate of approach to equilibrium. The second is known as the *transition matrix method*. While it involves slightly more overhead, it gives some limited dynamic information as well and can be used to evaluate possible move classes. Accordingly, we defer a discussion of the transition matrix method to the next section on dynamic information. While both techniques are more natural for discrete problems, which is how we phrase our descriptions, they can be adapted to continuum problems by binning the energies.

12.1.1 Histogram Method

The histogram or reweighting method [FS88, FS89] is a general method for estimating statistical averages. It combines the information obtained from equilibrium Monte Carlo simulations run at different values of certain parameters characterizing the system. Below we discuss only the application of this method to the problem of interest here: the calculation of the density of states, $\rho(E)$, from a sequence of constant temperature runs. The discussion uses the Lagrange multipliers technique for constrained minimization, a technique we met in Chapter 5.

Assume that M Monte Carlo simulations are performed at temperatures $T_1, T_2, \ldots, T_M$, yielding data streams of length $n_1, n_2, \ldots, n_M$. In the kth simulation of length n_k, we collect a histogram of the numbers of visits to states of each energy E. Let $N_k(E)$ be the number of visits to E during the kth simulation. Note that, for notational purposes, we use the inverse temperatures $\beta_1, \beta_2, \ldots, \beta_M$ throughout the rest of this section.

The model we are about to use for our estimation assumes that the data used in the model are uncorrelated, i.e., knowing the state at time t should tell us little about the state at time $t+1$. In fact, knowing the state at time t should tell us absolutely nothing about the next state, but we will settle for an approximation and accept small correlations. Let ξ_β be the correlation time of the walk at inverse temperature β, a quantity we discussed in Chapter 10. If we sample the states in our random walk every, e.g., $2\xi_\beta$ steps, the correlations between successive energy values become negligible. Assuming that the data were sampled in this fashion, we can consider $N_k(E)$ as the cumulated number of successes in a series of n_k independent trials. Each of these has two outcomes: either a state with energy E is visited (a success) or a state with some other energy $E' \neq E$ is visited (a failure). We assume that the probability of success, $\mathcal{P}(E)$, is small for each energy E and that the number of trials n_k is large. In this limit, the number of successes $N_k(E)$ is Poisson distributed [Fel66a]. Poisson distributed variables have the property that their mean and variance are equal. It thus follows that

$$\sigma^2_{N_k(E)} = \langle N_k(E) \rangle. \tag{12.5}$$

Another useful property is that the average of the histogram frequency $N_k(E)/n_k$ equals $\mathcal{P}(E)$, which again equals the equilibrium Boltzmann probability $\mathcal{P}_{\beta_k}(E)$ given in (12.1). Since then

$$\left\langle \frac{N_k(E)}{n_k} \right\rangle = \mathcal{P}_{\beta_k}(E), \tag{12.6}$$

we can write

$$\frac{N_k(E)}{n_k} = \rho_k(E) \frac{\exp(-\beta_k E)}{Z(\beta_k)}, \tag{12.7}$$

where $\rho_k(E)$ is the estimated density of states and $Z(\beta_k)$ the partition function. We treat $Z(\beta_k)$ as a number with zero variance, and for convenience we rewrite it in terms of its logarithm as

$$Z(\beta_k) = \exp(-\beta F(\beta_k)). \tag{12.8}$$

$F(\beta_k)$ is called the free energy at (inverse) temperature β_k and is also written in the shorter form F_k.

With this notation, we can solve (12.7) for $\rho_k(E)$ to get

$$\rho_k(E) = \frac{N_k(E)}{n_k} \exp(\beta_k(E - F_k)). \tag{12.9}$$

Taking the average of both sides gives

$$\rho(E) = \langle \rho_k(E) \rangle = \mathcal{P}_{\beta_k}(E) \exp(\beta_k(E - F_k)). \tag{12.10}$$

Note that the $F_k, k = 1, \ldots, M$ are so far unknown quantities. We proceed to find conditions that they satisfy.

The reweighted estimator $\rho(E)$ for the density of states is a weighted average of the ρ_k's involving a set of positive weights α_k which sum to one, $\sum_{k=1}^{M} \alpha_k = 1$:

$$\rho(E) = \sum_{k=1}^{M} \alpha_k \rho_k(E) = \sum_{k=1}^{M} \alpha_k \frac{N_k(E)}{n_k} \exp(\beta_k(E - F_k)). \tag{12.11}$$

We choose the weights α_k for our average by requiring them to minimize the variance $\sigma^2_{\rho(E)}$ of $\rho(E)$ and hence the statistical error in our estimate. By virtue of (12.5) and of the properties of the variance,[1] we find

$$\sigma^2_{\rho(E)} = \sum_{k=1}^{M} \alpha_k^2 n_k^{-1} \frac{\langle N_k(E) \rangle}{n_k} \exp(2\beta_k(E - F_k)). \tag{12.12}$$

By (12.6), we can substitute $\mathcal{P}_{\beta_k}(E)$ on the right-hand side of this equation for $\langle N_k \rangle / n_k$. Absorbing one of the exponential factors into $\mathcal{P}_{\beta_k}$ according to (12.10), we are left with minimizing

$$\sum_{k=1}^{M} \alpha_k^2 n_k^{-1} \rho(E) \exp(\beta_k(E - F_k)) \tag{12.13}$$

subject to

$$\sum_{k=1}^{M} \alpha_k = 1. \tag{12.14}$$

[1]Two properties of variance are used here: The variance of a sum is the sum of the variances; $\sigma^2_{A+B} = \sigma^2_A + \sigma^2_B$, and the variance of cA is c^2 times the variance of A; $\sigma^2_{cA} = c^2\sigma^2_A$ [Fel66a].

Equating the partial derivatives of the objective (12.13) to a Lagrange multiplier λ times the partial derivative of the constraint, we get for the jth partial derivative

$$2\alpha_j n_j^{-1} \rho(E) \exp(\beta_j (E - F_j)) = \lambda. \tag{12.15}$$

This equation can be solved for α_j

$$\alpha_j = \left(\frac{\lambda}{2\rho(E)}\right) n_j \exp(-\beta_j (E - F_j)). \tag{12.16}$$

Substituting these α_j's back into the constraint (12.14), we find

$$\left(\frac{\lambda}{2\rho(E)}\right) \sum_{j=1}^{M} n_j \exp(-\beta_j (E - F_j)) = 1. \tag{12.17}$$

Solving for the factor $(\frac{\lambda}{2\rho(E)})$ and substituting the solution into (12.16) gives

$$\alpha_j = \frac{n_j \exp(-\beta_j (E - F_j))}{\sum_{k=1}^{M} n_k \exp(-\beta_k (E - F_k))}. \tag{12.18}$$

Our next step is to construct a set of equations allowing us to find the unknown coefficients F_k. To this end, we pick an inverse temperature β_j and use (12.1) to express the partition function Z at this temperature in terms of the density of states ρ. This yields

$$Z(\beta_j) = \exp(-\beta F_j) = \sum_E \exp(-\beta_j E) \rho(E). \tag{12.19}$$

Inserting the reweighted estimate (12.11) into the above equation produces

$$\exp(-\beta_j F_j) = \sum_E \exp(-\beta_j E) \sum_{k=1}^{M} \frac{N_k(E)}{n_k} \alpha_k \exp(\beta_k (E - F_k)). \tag{12.20}$$

Equations (12.20) and (12.18) are $2M$ nonlinear equations for the $2M$ unknown quantities $\alpha_1, \alpha_2, \ldots,$ and $F_1, F_2, \ldots$. They can be solved iteratively as follows. Starting with some initial guesses for the F_k, we substitute these into (12.18) to calculate the values of the α_k. We then insert these values of the α_k and the F_k into the right-hand sides of (12.20). This then produces new values of the F_j on the left-hand side. This procedure is iterated until the changes in the F values from one iteration to the next are below some convenient tolerance.

This solution procedure must be repeated for each value of E. However, we can expect the F_k to change little from one E value to the next. Having obtained the solution for a given E, we can use it as the initial guess when solving for the next value, leading to convergence in just a few iterations. Once the F_k's are constructed for each E, the functional form of $\rho(E)$ and of all other equilibrium properties can be explicitly calculated.

To conclude this discussion, let us summarize the main properties of the histogram method, with special consideration given to its relevance for annealing. An annealing schedule usually goes through a series of temperature steps, which is also an essential element of the histogram method. Second, at each step we need to thermalize the system.

This is also true for the histogram method, where the interval between samplings must be of the order of the correlation time ξ_β of the Metropolis walk. It is possible to correct for the presence of correlations, but this requires an estimate of ξ_β. In any case, a careful evaluation of the density of states involves waiting longer at each temperature than one would usually be willing to do. However, the information gathered concerning the density of states for a class of problems can be usefully built into retrospective schedules.

12.2 Dynamic Information

The natural time scale ε mentioned in the previous chapter equals the inverse of the rate at which the mean energy relaxes to its equilibrium value $\langle E \rangle$. Unfortunately, in many systems with a complex energy landscape this final equilibration stage is not accessible to numerical experiments of realistic length for all temperatures lower than a certain glass temperature T_{glass}. At temperatures above T_{glass}, ε approximates the true relaxation time of the process. For $T < T_{\text{glass}}$, the true relaxation time becomes (by definition of T_{glass}) too large to be of computational interest and differs from the scale on which the energy is empirically observed to relax. In fact, several different scales of time become discernible below T_{glass}. For example, the equilibration of the standard deviation of the energy is significantly slower that that of the mean energy for a number of examples [RPS91].

The transition matrix method described in this section estimates the time scale on which the energy relaxes. Empirically, this time scale appears to reach a maximum value at a temperature T_{glass}, below which it stays approximately constant, as shown in Fig. 12.2. Below T_{glass}, the observed time scale ε has no relation to the real relaxation time of the problem.

12.2.1 Transition Matrix Method

The transition matrix method [AHM+88, RPS91] attempts to describe the relaxation of the system by considering the time evolution of the energy values. The set of possible energies is binned (if not already discrete) and energy-to-energy transition probabilities are estimated by maintaining an array of counters Q_{ij} of attempted moves. Every time a move is attempted from energy E_j to energy E_i, counter Q_{ij} is incremented by one. At the end (or along the way) we get an estimate of the (infinite temperature) energy-to-energy transition matrix P by normalizing the columns

$$P_{ij} = \frac{Q_{ij}}{\sum_{i'} Q_{i'j}}. \tag{12.21}$$

After normalization, the entries P_{ij} are estimates of the probability of transition[2] between energy bins j and i, at infinite T. The dominant eigenvector of P, corresponding to the eigenvalue 1, is proportional to the equilibrium probability distribution, which, by virtue of (12.1), is simply the density of states $\rho(E)$.

[2]Note, however, that the sampled energy range strongly depends on the temperature. To restrict to the relevant energy range, masking is usually employed in conjunction with this technique, i.e., only a portion of this matrix is used. To preserve detailed balance, any transitions that would jump to regions outside the masked portion of the matrix have to be treated as rejected moves, i.e., the diagonal of the masked matrix must be adjusted to keep the column sums equal to one.

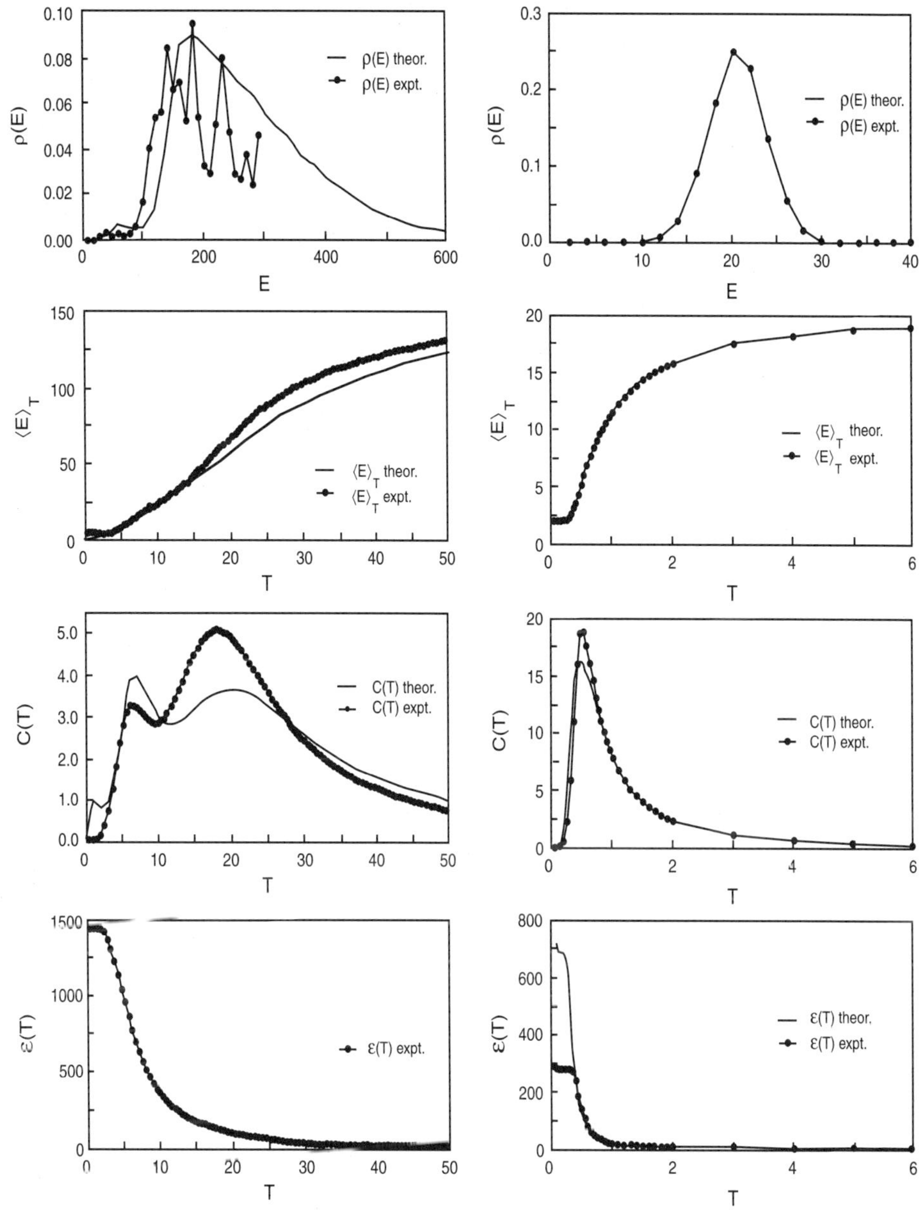

Figure 12.2. *Two empirically obtained thermodynamic portraits* [AHM+88]. *The left column shows the data for a seismic deconvolution problem, while the right column shows the data for a graph bipartitioning problem. The quantities shown are the density of states $\rho(E)$ (first row), the mean energy $\langle E \rangle_T$ (second row), the heat capacity $C(T)$ (third row), and the relaxation time $\varepsilon(T)$ (fourth row). Theoretical predictions are also shown where available.*

To find the finite temperature counterpart, we "Boltzmannize" P by defining

$$M_{i,j}(T) = \begin{cases} P_{ij} & \text{if } E_i \leq E_j, i \neq j, \\ P_{ij}\exp(-\Delta E/T) & \text{if } E(i) > E(j), \\ 1 - \sum_{k \neq j} M_{k,j}(T) & \text{if } i = j. \end{cases} \tag{12.22}$$

In that case, the Boltzmann distribution $\mathcal{P}_T(E_i)$ is proportional to the dominant eigenvector of $M(T)$. The time scale $\varepsilon(T)$ is estimated by $-1/\ln(|\lambda_2|)$, where λ_2 is the second-largest eigenvalue of $M(T)$. (The largest eigenvalue is always equal to 1, as discussed in Chapter 6.)

We can use M as a way to simulate the energy-lumped dynamics of the process. As shown in the Appendix to this chapter, the true energy-lumped transition matrix is exact in its prediction of equilibrium quantities. This does not mean that a runtime estimate of this matrix will predict $\rho(E)$ perfectly; it does mean that energy lumping does not introduce any bias into this way of calculating $\rho(E)$. On the other hand, the method has certain inherent inaccuracies in how it estimates nonequilibrium quantities. Nevertheless, the estimates it provides can be useful for adaptive schedules, as described in the next chapter. Furthermore, having the infinite temperature transition probability matrix P can also give us an estimate of the combinatorial work (see section 10.1.4) of equilibrating to the next temperature. This combinatorial work can be used to judge the quality of a move class and is given by

$$K(\mathcal{P}_{T+\Delta T} | P \cdot \mathcal{P}_T), \tag{12.23}$$

where K is the Kullback information (see footnote 2, section 10.1.4).

12.3 Time-Resolved Information

The ensemble size optimization (see Chapter 8) as well as the estimation of the global minimum (see Chapter 14) needs empirically sampled distributions of the best energy seen, E_{bsf}, as a function of time during the random walk. Empirical estimates of these distributions can be collected using an ensemble, provided we store the value of the best energy seen by each random walker. Typical implementations collect only rather coarse distributions of E_{bsf} by binning the values into something on the order of 20 categories. At certain selected times along the way, the algorithm needs to pause, collect the E_{bsf} values from the random walkers, and count the frequencies of E_{bsf} falling in each of the bins. The collection times selected vary greatly with the problem but typically are done at exponentially increasing intervals, e.g., $t = 100, 200, 400, 800, \ldots,$ for convenience in the calculations. Note that these distributions depend on the details of the process, such as the choices of cooling schedule and move class used along the way.

12.A Appendix: Why Lumping Preserves the Stationary Distribution

The energy-to-energy transition matrices estimated in section 12.2.1 represent a sort of shadow of the full process that takes place on the much larger set of states. The method whereby this shadow is cast is known as *lumping*. In terms of the states of the original underlying chain, the lumped chain is defined by partitioning the states of the original chain into subsets that represent states of the lumped process. The transition probabilities into a lumped state are obtained by summing the probabilities of transitioning to the states included in the lumped state. Transition probabilities out of the lumped state on the other hand are obtained by averaging the outward bound transition probabilities. This is illustrated in the following example.

EXAMPLE 12.1. Consider the two-basin Example 6.9. Recall that the transition matrix for this example is given by (6.49) as

$$M = \begin{bmatrix} .9 & .2 & 0 & 0 \\ .1 & .79 & .04 & 0 \\ 0 & .01 & .66 & .1 \\ 0 & 0 & .3 & .9 \end{bmatrix}. \tag{12.A.1}$$

Let us define the lumped state $A = \{1, 2\}$ and stipulate that a transition to A occurs if a transition to any state of A occurs. Thus the transition probabilities

$$\hat{M}_{A3} = M_{13} + M_{23} = 0.04 \tag{12.A.2}$$

and

$$\hat{M}_{A4} = M_{14} + M_{24} = 0 \tag{12.A.3}$$

are just the sums of the probabilities in the unlumped chain. The transition probabilities out of the lumped state, M_{3A} and M_{4A}, depend on a knowledge of how often we are in state 1 or state 2 when the lumped process is in A. To get an appropriate description of the lumped process, we need to average the outward bound probabilities and, to this end, we need to know the weights with which to average. Recalling that the stationary distribution of this chain is $[\frac{1}{2}, \frac{1}{4}, \frac{1}{16}, \frac{3}{16}]^t$, we see that at equilibrium the chain finds itself in state 1 twice as often as it finds itself in state 2. We choose these weights for calculating the average:

$$\hat{M}_{3A} = \frac{2}{3}M_{31} + \frac{1}{3}M_{32} = 0.003\bar{3} \tag{12.A.4}$$

and

$$\hat{M}_{4A} = \frac{2}{3}M_{41} + \frac{1}{3}M_{42} = 0. \tag{12.A.5}$$

This gives the lumped transition matrix

$$\hat{M} = \begin{bmatrix} .996\bar{6} & .04 & 0 \\ .003\bar{3} & .66 & .1 \\ 0 & .3 & .9 \end{bmatrix}. \tag{12.A.6}$$

The operations carried out on the original matrix M to give the lumped matrix $\hat{M}$ correspond to summing the rows and averaging the columns in each lump. This can be expressed in terms of matrix operations as

$$\hat{M} = U \cdot M \cdot V, \tag{12.A.7}$$

where U is a block matrix of 0's and 1's and V is a block matrix of the weights to be used in the averaging [KS60].

EXAMPLE 12.2. For the example above,

$$U = \begin{bmatrix} 1 & 1 & 0 & 0 \\ 0 & 0 & 1 & 0 \\ 0 & 0 & 0 & 1 \end{bmatrix} \tag{12.A.8}$$

and

$$V = \begin{bmatrix} \frac{1}{3} & 0 & 0 \\ \frac{2}{3} & 0 & 0 \\ 0 & 1 & 0 \\ 0 & 0 & 1 \end{bmatrix}. \tag{12.A.9}$$

Note that the portions of U and V corresponding to the unlumped states is just an identity matrix.

That there is an arbitrariness of choosing the weights for the averaging of outward bound transition probabilities hints at a problem regarding the accuracy of the lumped process as a representation of the real underlying dynamics.[3] It is a theorem [KS60] that if we average the outward bound transition probabilities using the stationary distribution of the chain (suitably renormalized) then the lumped process correctly reproduces the stationary behavior. In particular, the stationary distribution of the lumped chain is the lumped stationary distribution of the original chain [AHM+88]. The implication for us is that the stationary distribution of the energy lumped chain is indeed the Boltzmann distribution $\mathcal{P}_T$.

EXAMPLE 12.3. Continuing our example above, we note that when states 1 and 2 are lumped together, the stationary distribution of M, $[\frac{1}{2}, \frac{1}{4}, \frac{1}{16}, \frac{3}{16}]^t$, becomes the stationary distribution of $\hat{M}$, $[\frac{3}{4}, \frac{1}{16}, \frac{3}{16}]^t$.

While the matrix of the lumped chain can give perfectly accurate information regarding the stationary distribution, the relaxation time of the lumped chain in general underestimates the relaxation time of the original chain. For the example above, $\varepsilon(M) = 82$ while $\varepsilon(\hat{M}) = 80$. Despite this, the relaxation time of the energy lumped simulated annealing problem does give a useful time scale for controlling the cooling process.

[3]In general, the real process viewed at the level of the lumps is not even Markovian [KS60]. Here we have defined an approximate description which is Markovian.

Chapter 13
Selecting the Schedule

The schedule for an annealing problem is usually a decreasing sequence of temperatures $T(n)$ successively utilized in the transition probabilities of the algorithm. In many cases the performance of annealing appears to be rather insensitive to the form of $T(n)$ [JAMS89, JAMS91, JM97, SNR+88]. Given this fact many practitioners opt for simplicity, and the number-one choice is the exponential cooling schedule described in the original papers by Kirkpatrick et al. [KJV83] and Černý [Čer85] and further popularized by a number of authors [PFTV81, vLA87, Aze92]. There exist problems for which the choice of schedules makes a significant difference. For seismic deconvolution, for example, the reported average improvement is a factor of two [MV91] over a carefully optimized exponential schedule. On the other hand, the improvement reported for graph bipartitioning and the traveling salesman problem is very modest [SNR+88, Rup88]. Once a good schedule has been found, it tends to behave very similarly for different instances within a class of problems, and a generic fitted schedule can be implemented.

Most schedules leave the user with a free parameter that determines the overall rate of cooling. Choosing this rate to be a constant introduces a time scale into the algorithm that must be matched with the natural time scales of the problem. Glassy systems have a multitude of relaxational time scales spanning many orders of magnitude. A rapid cooling tends to quench the system's degrees of freedom corresponding to the slow time scales, while a slower cooling allows more of these to come into play and more of the configuration space to be explored. Thus, the slower we cool, the better the final optimization result can be expected to be. For example, one study [GSL86b] found a logarithmic dependence of the lowest energy of a spin glass model as a function of the cooling rate. While this leaves the overall question of how fast to cool up to our computer budget and eventually to how much we care for the quality of the solution, the possibility of cooling at a nonuniform rate, or even of intercalating cooling and reheating, is open to optimization. This is exactly where more elaborate schedules can give significant improvement.

Whatever schedule one uses, the start and stop criteria need to be specified. How this can be done directly by using natural scales for the problem is the topic of the first section. Section 13.2 discusses simple cooling schedules, including exponential cooling. Section 13.3 discusses several adaptive cooling schedules that again exploit the natural time and energy scales in a problem.

13.1 Start and Stop Temperatures

The starting and ending temperatures of the algorithm, T_0 and T_{final}, are usually set by using the natural scales of the problem. White [Whi84] advocates choosing T_0 so that the system should initially be sufficiently hot to allow large fluctuation in the energy. The scale of these fluctuations is given by the standard deviation of the energy at infinite temperature: $\sigma_{E|_{T=\infty}} = \sqrt{\langle E^2\rangle_{T=\infty} - \langle E\rangle^2_{T=\infty}}$. If not known a priori, this quantity can be estimated with modest effort by randomly sampling the configuration space. The same procedure will also provide a good random starting configuration. In practice one should choose T_0 moderately larger than σ_E, e.g., twice as large. In ensemble-based approaches one needs a set of random initial configurations rather than just one. This is also easily achieved by saving an appropriate number of configurations obtained in the random sampling.

A simpler criterion is to set T_0 so most moves are accepted initially. If one parametrizes the temperatures using the reciprocal temperature $\beta = 1/T$, it is perfectly reasonable to start with $\beta = 0$, which means that all moves will be accepted. A more popular version is to select T_0 so that exactly half the moves will be accepted. In any case, the issue is not critical as relaxation is very fast at high temperatures.

The stop temperature T_{final} is usually set adaptively by specifying a number of steps N_{stop} such that, if the energy does not change in the last N_{stop} steps, then it is time to stop the algorithm. For discrete problems, White [Whi84] suggests taking the stop temperature of the order of the smallest energy scale in the system, which he takes to be the smallest possible change in the energy ΔE during a single move.

13.2 Simple Schedules

The simple schedules described below depend on only one or two parameters and have a functional dependence on time that can usually be expressed in closed form in terms of the start and stop temperatures and the total number of steps.

13.2.1 The Sure-to-Get-You-There Schedule

The issue of schedules has both theoretical and practical aspects. A well-known result in the former category is due to Geman and Geman [GG84], who introduced a schedule that guarantees convergence to the optimal solution. Letting E_{bsf} be the best energy seen during the walk of length t, the schedule has the property [GG84, Haj88] that

$$\lim_{t\to\infty} \text{prob}\{E_{\text{bsf}} = E_{\text{minimum}}\} = 1. \tag{13.1}$$

The schedule itself is given by

$$T(t) = \Delta E^{\max}_{\text{activation}} / \ln(t+1), \tag{13.2}$$

where $\Delta E^{\max}_{\text{activation}}$ is the largest activation energy in the problem, i.e., the largest energy difference that must be overcome along the paths leading out from the bottom of any sub-optimal basin (see section 6.4.2). It is interesting that, with this schedule, one can move sufficiently slowly to drain all of the probability from every basin into the global minimum.

It can further be shown [Aze92] that the probability of not finding the minimum decays with time as $P(E_{\text{bsf}} > E_{\text{min}}) \propto t^{-x}$, where x is a suitable exponent. Unfortunately, it is also clear that with such slow logarithmic decay of the temperature and a large $\Delta E^{\text{max}}_{\text{activation}}$, the time required for the annealing to stop is truly enormous and of little relevance to real problems.

Several practitioners of simulated annealing use a schedule that has the form (13.2) but with a smaller constant in the numerator. Such a schedule

$$T(t) = d/\ln(t+1) \tag{13.3}$$

will (eventually) drain the probability out of all basins with depth less than d. In combination with the no-rejection Metropolis schemes discussed at the end of Chapter 10, the above schedule is a viable option. The quantity d must then be chosen with an eye to the total length t_{max} of the simulation one is able to afford, i.e., such that $T(t_{\text{max}})$ is small compared to the typical energy change required by a move.

13.2.2 The Exponential Schedule

An exponential schedule (sometimes also referred to as a geometrical schedule) has the form

$$T(t) = T_0 \alpha^t. \tag{13.4}$$

As mentioned above, this is by far the most commonly used schedule. The cooling factor α is usually set to a number close to 1, say 0.99. The temperature is typically updated only every k moves with k on the order of 10 to 100, making the actual functional form

$$T(t) = T_0 \alpha_2^{\lfloor t/k \rfloor}. \tag{13.5}$$

This has the effect of allowing partial equilibration at each temperature and corresponds to a step function approximation to the exponential function in (13.4) with $\alpha = \alpha_2^{1/k}$. Note that if T_0, T_{final}, and t_{final} are known, one can solve for α from (13.4).

13.2.3 Other Simple Schedules

Many other simple functional forms have been employed for the cooling schedule. The most noteworthy among these is increasing $\beta = 1/T$ linearly as a function of time [MR81, SH87]:

$$\beta_{t+1} = \beta_t + m, \tag{13.6}$$

where m is a positive constant. Like α in (13.4), m can be determined from the value of T_0, T_{final}, and t_{final}. Several authors have reported good performance with this schedule [SH87, SNR+88]. With about the same effort, one can also decrease T linearly in time or as an inverse power $T_0 t^{-x}$ with almost any $x > 0$, but we are not aware of any claims in the literature for particular benefits of doing this.

13.3 Adaptive Cooling

Failing rigorous proofs on the benefits of more involved approaches, we once again resort to the folk theorem: The more one uses the structure of the problem, the better. Adaptive cooling schemes collect information concerning the system and use it to make educated guesses about the way the temperature should be changed.

An interesting adaptive cooling schedule that simultaneously adapts the neighborhood size (move class) was introduced by Morey, Scales, and Van Vleck [Morey98]. They used the lumped transition matrix method outlined in the previous chapter to estimate a mean first passage time to the ground state. Their estimate depends on the temperature and the neighborhood size and they pause every few hundred steps to vary these parameters slightly from their current values to minimize their estimate of this expected hitting time.

Adaptive schemes can be implemented on the fly or *retrospectively*. Retrospective implementations use data collected during previous runs to calculate the desired schedule $T(t)$. Retrospective data analysis is popular for several reasons. Generally it tends to be more stable than runtime implementations which have to rely on statistical estimators that may at times be overwhelmed by noise. Using a retrospective analysis also enables the algorithm to run embarrassingly parallel since no runtime communication is needed.

Recall (Chapter 3) that $\langle\langle E\rangle\rangle_t$ stands for the instantaneous average at time t over a finite ensemble of walkers, while $\langle E\rangle_T$ stands for the equilibrium average at temperature T, which is evaluated using the Boltzmann distribution. We denote the temperature dependence of this average by a subscript, which we often omit in an attempt to keep the notation uncluttered. The equilibrium standard deviation of the energy is denoted by $\sigma_E(T)$, where again the T dependence is often suppressed. We do not introduce a notation for the instantaneous standard deviation of the ensemble energy. All the formulas we use call for the equilibrium average. Despite this, the $\sigma_E(T)$ used in the implementations is often based on an ensemble average, a time series average, or some combination to make up for a small ensemble size. In these implementations, $\sigma_E(T)$ is estimated by $\sqrt{\langle\langle E^2\rangle\rangle_t - \langle\langle E\rangle\rangle_t^2}$ at the end (or along the last stretch) of a fixed temperature run.

13.3.1 Using the System's Scale of Time

Hoffmann et al. [HWdGH91] suggested an appealing cooling schedule which is the simplest to implement among our adaptive schedules. Their schedule relies on the statistical behavior of an ensemble of random walkers and is implemented as follows. The temperature is always decreased by a fixed factor α, but the time t spent at each temperature T is allowed to vary in an adaptive way. Ideally, one would like to equilibrate the system at each temperature. Suppose that, at the nth cooling step, the temperature is lowered from T_n to $T_{n+1} = \alpha T_n$. Since the heat capacity $C(T)$ is positive, the system responds by lowering its ensemble mean energy $\langle\langle E\rangle\rangle_{n+1}(t)$ toward its equilibrium value at T_{n+1}. In an infinite ensemble, the ensemble mean would approach the new equilibrium value $\langle E\rangle_{T_{n+1}}$. In a finite ensemble of the kind used in simulations, however, the mean fluctuates in time around its fixed equilibrium value. A fluctuation that leads to an increase in the ensemble average energy is taken by Hoffmann et al. [HWdGH91] as an indication that the correlation function has decayed. This implies (see section 6.4) that the ensemble is sufficiently close to thermal equilibrium and therefore the temperature can be further lowered. While these authors report good results for a 532-city traveling salesman problem, their procedure is slightly flawed since it depends on the size of the ensemble used. A better procedure is to wait until a positive fluctuation of the ensemble average energy $\langle\langle E\rangle\rangle$ is at least of size $c\sigma_E/\sqrt{N}$, where c is a preset constant, σ_E the standard deviation of the energy in the ensemble, and N is the ensemble size. In summary, the criterion is to test for each time t whether

$$\langle\langle E\rangle\rangle_{t+1} - \langle\langle E\rangle\rangle_t > c\sigma_E/\sqrt{N} \tag{13.7}$$

and to lower the temperature if condition (13.7) is satisfied. We call this approach the *wait-for-a-fluctuation* schedule. Note that this schedule uses the system's internal time scales at different temperatures and decreases the temperature nonlinearly in time.

13.3.2 Using the System's Scale of Energy

The schedule above used the system's own time scales. Our next schedule uses the system's own local scale of energy as measured by σ_E [SSV85, HRSV86]. This cooling schedule is available as part of the Timberwolf package developed at UC Berkeley. It often goes by the name of *constant speed* cooling since it adjusts the cooling rate so that

$$\frac{\frac{d\langle E\rangle}{\sigma_E}}{dt} = \frac{1}{\sigma_E}\frac{d\langle E\rangle_{T(t)}}{dt} = -c, \tag{13.8}$$

where c is a small positive constant. Using the fact that $\frac{d\langle E\rangle}{dT} = \sigma_E^2/T^2$, this determines the rate at which to cool, $\frac{dT}{dt}$, as

$$\frac{dT}{dt} = -cT^2/\sigma_E. \tag{13.9}$$

In summary, to implement this algorithm we update the temperature T using

$$T_{n+1} = T_n - cT^2/\sigma_E \tag{13.10}$$

and take a fixed number of time steps between updates. The constant c is the "how much we care" parameter discussed in the introduction to this chapter.

13.3.3 Using Both Energy and Time Scales

The schedule that combines the benefits of the previous two schedules is known as *constant thermodynamic speed* annealing. This schedule uses both the system's time and energy scales. The simplest (and arguably the best) implementation is to increment the temperature using (13.10) combined with the wait-for-a-fluctuation update rule in (13.7).

Fig. 13.1 illustrates a constant thermodynamic speed process [Ped90]: A system (the horse) is coaxed along a sequence of states by keeping the temperature of the heat bath with which it is trying to equilibrate (the carrot) a fixed distance ahead. This is called the "horse–carrot process."

For the control of thermodynamic processes, the theorem that corresponds to the cartoon is known as the horse–carrot theorem. This theorem has two versions: discrete

Figure 13.1. *The horse–carrot caricature of a constant thermodynamic speed annealing process.*[1]

[NSAA85] and continuous [SB83]. In the discrete version, the control temperature is changed in sizeable increments, each followed by a relaxation nearly reaching equilibrium. This is what the combination of (13.10) and (13.7) achieve. The continuous version changes the control temperature as nearly continuously as convenience allows, while always keeping the cold reservoir (whose contact the control temperature simulates) a fixed thermodynamic distance ahead of the state of our ensemble. To achieve this, we need a direct prescription for continuously decreasing the temperature T. If we let ε represent the time scale for the approach of $\langle E \rangle$ to its equilibrium value, cooling with the system's own scale of time amounts to spending a certain fixed number of ε's at each temperature. Thus the effect of the algorithm is to keep

$$\frac{\frac{d\langle E\rangle}{\sigma_E}}{\frac{dt}{\varepsilon}} = \frac{\varepsilon}{\sigma_E}\frac{d\langle E\rangle}{dt} = -v, \tag{13.11}$$

where v is a small positive constant known as the thermodynamic speed. Translating this to dT/dt, we get

$$\frac{dT}{dt} = -v\frac{T^2}{\varepsilon\sigma_E}. \tag{13.12}$$

This implementation makes explicit use of the relaxation time. For an implementation without the need to estimate a time scale [RPS91], we can alternatively keep adjusting T so

$$\frac{\langle\langle E\rangle\rangle_t - \langle E\rangle_T}{\sigma_E(T)} = -v. \tag{13.13}$$

Fig. 13.2 illustrates the improvement found for a seismic deconvolution problem using a constant thermodynamic speed schedule. This adaptive schedule has proved to be more

[1]Figure from [Ped90] with the permission of Jacob Mørch Pedersen and Lene Mørch Pedersen.

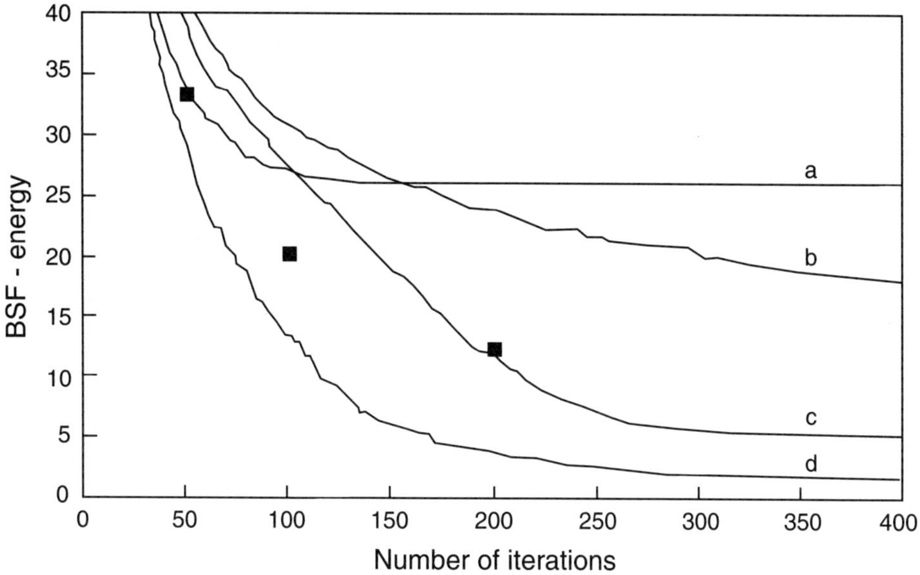

Figure 13.2. *The effect of schedules on performance. The data shown are redrawn from* [Ped90] *and show the effect of different schedules on the mean BSF energy averaged over* 350 *runs. The problem considered is seismic deconvolution—Problem* A *in Chapter* 2. *The methods compared are* (a) *quench,* (b) *random search,* (c) *annealing with an exponential schedule, and* (d) *constant thermodynamic speed annealing. The squares indicate the performance of exponential annealing schedules, empirically tuned to* 50, 100, *and* 200 *iterations, respectively.*

popular as a retrospective schedule, i.e., a schedule in which information about past runs is stored and used on similar problems.

Constant thermodynamic speed first arose as a control strategy for minimizing the entropy produced in a thermodynamic process [SB83, NSAA85]. Given how many relaxations a system is to undergo, proceeding at constant thermodynamic speed minimizes the total entropy production in the process. The relevance of this for global optimization hinges on a (questionable) relation between entropy production and the combinatorial work discussed in Chapters 10 and 12 [ZS92, SHHN88]. This combinatorial work has an information theoretic interpretation as the number of extra questions asked (function values evaluated) in selecting out states of the system that are representative of the next stationary distribution. By this interpretation, minimizing the total entropy production minimizes the number of function evaluations needed to select the sequence of ensemble states $\mathcal{P}_{T_i}$.

The above, tentative interpretation notwithstanding, there is a completely different sense in which this schedule is optimal: It cools as fast as possible consistent with the constraint of never being further than a certain distance away from equilibrium [NS88]. For nearby states, it turns out that the differential length $\frac{|dE|}{\sigma_E}$ is also proportional to the χ^2 statistic. Thus, in a finite ensemble context, being within a certain distance of the stationary distribution means that our ensemble, to a certain level of statistical confidence measured by a χ^2 test, is a sample taken from the Boltzmann distribution. The additional fact leading to optimality is that the lower the temperature used by the Metropolis algorithm, the faster the cooling rate $d\langle E\rangle/dt$. The optimality follows [NS88].

Constant thermodynamic speed schedules have been tested on a number of examples [SNR+88, Rup88, RPS91, MV91, VM91, YBS91, Ped90, NA98]. While it has consistently outperformed other schedules, the margin by which it improved performance has been disappointingly small in several of the examples. Our conjecture is that, given only thermodynamic information, constant thermodynamic speed cooling is optimal.[2]

13.4 Nonmonotonic Schedules

So far, we have discussed schedules where the temperature decreases monotonically in time. Several approaches have recently been developed where this constraint is relaxed and where the system undergoes—in one way or another—a series of thermal cycles. This section describes two such approaches.

Schneider, Morgenstern, and Springer [SMS98] introduced *bouncing* in connection with a study of a specific traveling salesman problem. After an initial cooling, the system is repeatedly heated to a certain temperature T_b and then cooled anew. Two versions of cooling with an exponential schedule were implemented: one using Metropolis and one using threshold accepting (see section 11.2). There are different ways to choose T_b. The simplest is to pick T_b slightly below the temperature T_c at which the heat capacity $C = \frac{d\langle E\rangle}{dT}$ has its maximum. At $T = T_c$ one has the largest energy change per unit change in temperature. Below T_c the system is partly frozen, that is, many edges connecting neighbor points of the tours will, with overwhelming probability, not change any more during the time scale of the simulation. Schneider et al. refer to these edges as the backbone of the problem.

In a slightly more complicated version of bouncing, the maximum temperature $T_b(i)$ used in the ith cooling–reheating cycle is allowed to decrease as i increases. This introduces a secondary cooling schedule, which should not be confused with the primary schedule used in the cooling part of each stage. Schneider et al. used an ensemble of independent bouncers to adaptively determine how $T_b(i)$ should be decreased: If the ensemble average of the energy at the end of cooling stage $i+1$, $\langle\langle E\rangle\rangle_{i+1}$ satisfies

$$\langle\langle E\rangle\rangle_{i+1} \geq \langle\langle E\rangle\rangle_i, \tag{13.14}$$

then the bouncing temperature for the next stage is decreased by a constant factor of 0.99. Note that T_b is not decreased after each cycle but only following those cycles satisfying (13.14).

A method called *thermal cycling* by Möbius et al. [MNDS+97] is similar to bouncing in that it iterates cycles consisting of a search for a local minimum followed by a partial randomization of the configuration found. Randomization at cycle i is achieved by performing a fixed number of Metropolis steps at temperature T_i. Unlike bouncing, the search for

[2] Given more complete information about a problem, e.g., its complete state-to-state transition probability matrix, a different cooling schedule does better. This has been demonstrated for toy examples of the type described in Example 6.10 for which an optimal cooling schedule can be analytically calculated [HS90]. This schedule does not coincide with constant thermodynamic speed, although it is based on much more detailed information about the system.

the local minimum utilizes a greedy local search method, i.e., a quench. The local minima found are archived, and the randomization is repeated. If the algorithm fails to improve the best configuration stored in the archive, the temperature T_i is reduced by a fixed factor. The method was tested on different standard instances of the traveling salesman problem, leading to considerable improvement over standard simulated annealing.

Intuitively, the success of nonmonotonic schedules stems from the ability to escape the valley of the local energy minimum found during cooling while still retaining the information about the good parts of the solution, which, for traveling salesman problems, is encoded in the frozen edges of the tours. Each new cooling stage or local search explores a region of configuration space that is close, in some sense, to the region explored in the previous attempt. Viewed in this way, these methods resemble basin hopping. As in basin hopping, large moves that destroy the backbone are undesirable. Indeed, when large changes are induced by choosing $T_b > T_c$, bouncing does not lead to any improvement as compared to a single cooling stage.

13.5 Conclusions Regarding Schedules

We discussed a number of different approaches, starting with the simplest exponential schedules and gradually moving to more complex ones, which include adaptive, ensemble-based elements and a nonmonotonic temperature profile. Not surprisingly, these later developments offer considerable improvements over the simpler approaches. Which schedule is best for a given problem depends on the amount of development time one is willing to invest versus the computing effort that is expected to follow after the development phase is finished. Implementing anything but simple schedules is probably more hassle than a casual practitioner is willing to put up with. On the other hand, for large-scale applications, implementing an adaptive schedule is usually well worth the extra development time.

Chapter 14

Estimating the Global Minimum Energy

In this chapter we describe a heuristic way to estimate the energy of the ground state of an optimization problem. The method was introduced by Sibani et al. [SPHS90], who analyzed in this fashion a traveling salesman problem and a graph bipartitioning problem. Later, it was more fully tested by Tafelmayer and Hoffmann [TH95]. The method provides estimates of the ground state energy in a relatively inexpensive way but does not give any information regarding the corresponding configuration. The estimate can be used to provide a stopping criterion for the annealing procedure.

The approach rests on the assumption that a characteristic geometrical structure of state space exists, which can be at least partially uncovered by statistical data sampling. The fact that it works reasonably well for a number of systems in itself poses an interesting theoretical puzzle. This issue of state space structure and its bearing on optimization has intrinsic interest and will be further discussed in Chapter 15. Here we are mainly concerned with a straightforward description of how the method is implemented.

Consider an (ideally infinite) ensemble of walkers, each maintaining a record of the lowest energy E_{bsf} they have seen during the walk. Conforming to the nomenclature of the rest of the book, we refer to E_{bsf} as the best-so-far energy. At each point of time, the ensemble generates a distribution of best-so-far energies, $F(t, E_{\text{bsf}})$, whose properties are the object of our present analysis. The main conjecture is that in a nonempty class of problems, the best-so-far distribution obeys specific scaling relations, provided that the zero of the energy axis is chosen to be the ground state energy.

Specifically, the ansatz of Sibani et al. [SPHS90] states that if E_g is the true ground state energy of the problem, then the kth moments defined by

$$\mu_k(t) = \langle (E_{\text{bsf}}(t) - E_g)^k \rangle (t) \tag{14.1}$$

fulfill

$$\mu_k(t) \cdot (1 + (t/t_0)^\alpha)^k = C, \tag{14.2}$$

where $\langle\rangle$ indicates, as usual, the ensemble average, the exponent α fulfills $0 < \alpha < 1$, and t_0 and C are constants. A proof of 14.2, or even a good a priori characterization of the systems to which it applies, is currently lacking.

To apply the formula, one starts with an ensemble of randomly initialized random walkers that explore the system configuration space with a simulated annealing algorithm or even a Metropolis walk at constant temperature. In a certain time interval $[0, \tau]$, the $E_{\rm bsf}$ distribution and a set of its moments μ_n are calculated for a number of different n values. In [SPHS90], values of n from -4 through 4 were used.

To check how well the scaling relation (14.2) is fulfilled given a putative ground state energy E_g^p, this value is inserted in the left-hand side of the ansatz, which is then fitted to a constant. Finally, the root mean square deviation of the data points from this fitted constant is extracted. This root mean square value is a figure of merit, $Q(E_g^p)$, which, as stressed by the notation itself, depends on the value of the putative ground state energy chosen in the first place.

The fitting procedure is now repeated for a set of different E_g^p, yielding the functional dependence of $Q(E_g^p)$ on its argument. Interestingly, in the examples considered by Sibani et al., the function $Q(E_g^p)$ has a very sharp minimum, i.e., the function values change by several orders of magnitude for a modest variation of E_g^p around a certain value E_g^e, which is then used as an estimate of the the ground state energy of the problem at hand.

In a special case with known ground state energy (a traveling salesman problem with the cities placed on a regular grid), the estimate turned out to be a few percent from the

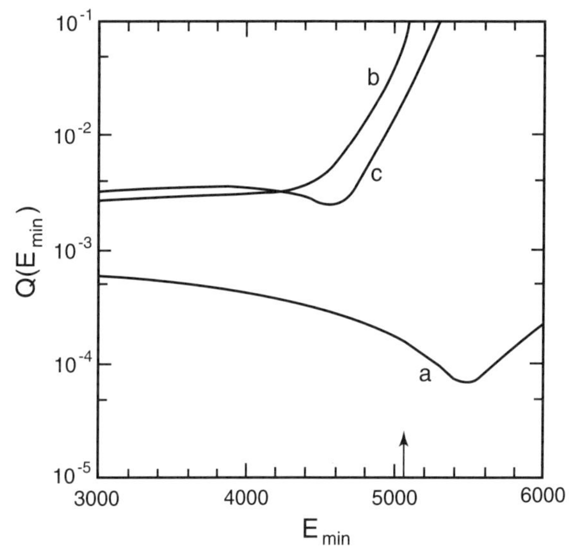

Figure 14.1. *Estimating the ground state energy. The data shown are redrawn from* [TH95] *and illustrate several estimates of the ground state energy whose actual value is shown by the arrow. The example used is the Grötschel traveling salesman problem. The figure of merit described in the text is plotted versus the putative global minimum of the problem. Curve* a *is calculated using simulated annealing and the wait-for-a-fluctuation schedule described in section* 13.3.1. *Curve* b *is obtain by threshold acceptance with a constant temperature, while curve* c *stems from a constant temperature walk. The minimum of curve* c *is about* 10% *off the true minimum of the problem, which is shown by the arrow.*

actual value. The estimate will in general depend somewhat on the time window in which the calculation is performed, and it will actually improve as the time window is extended. However, rather short walks already seem to yield reasonable estimates.

Tafelmayer and Hoffmann [TH95] made an empirical study of the applicability of the ansatz to larger-scale problems. They considered the Grötschel drilling problem, which is a traveling salesman problem in which a drill has to travel to 442 different drill holes, and a mean-field spin glass problem with 1024 spins, any two spins coupled by a random Gaussian distributed coupling constant with zero average and unit variance.

Both problems were studied with a host of different approaches based on simulated annealing and threshold accepting. The threshold accepting method did not in general lead to a scaling behavior of the moments and to a pronounced minimum for the figure of merit. This was, however, the case for the simulated annealing algorithm in combination with an adaptive schedule. It was also the case for constant temperature walks. As shown in Fig. 14.1, the predictions obtained for the ground state were then within 10% of the lowest energy on all the examples studied.

In conclusion, the scaling ansatz and the estimates obtained have, as one would expect, only approximate validity. Their applicability can be checked in each case at a very modest computational overhead. When the method works, it can provide a useful stopping criterion for annealing. Additionally, the existence of reasonably well-defined scaling behavior demonstrates the existence of geometrical structure in the configuration space of hard combinatorial problems.

Part IV

Toward Structure Theory and Real Understanding

Part IV continues our development of the physical understanding one can gain about annealing algorithms. The theory developed here sheds a great deal of light on the methods presented in Part III and is, we believe, indispensable for the ultimate understanding that can bring this subject from the realm of heuristics to the realm of provably optimal algorithms. This latter realm is likely to divide the family of global optimization problems into classes based on structural features of the problem and the best approach is likely to depend on class. This motivates our belief that the important next step for global optimization will require a better understanding of the geometrical structure of complex landscapes. Chapter 15 discusses how this structure can be analyzed, emphasizing the issues of coarse graining and of how the multitude of traps invariably present in such landscapes should be characterized. In our discussion we use a mixture of simple arguments based on toy models and numerical evidence pertaining to real examples. The Appendix to Chapter 15 is meant to convey some intuition regarding entropic barriers, an issue of considerable importance in high-dimensional problems. Chapter 16 uses the concepts and results discussed in Chapter 15 to ponder the puzzling question of why annealing works at all.

Chapter 15

Structure Theory of Complex Systems

Recall that simulated annealing is a sequence of thermalization processes in the state space of the problem, with a fictitious temperature determining the likelihood of uphill moves. If a system of this type remains equilibrated while its temperature is progressively lowered, the Boltzmann probability distribution is shifted toward low-energy states and eventually becomes nonzero only for the lowest energy configuration—the object of the search. Unfortunately, as is well known to practitioners of simulated annealing, it is generally impossible to thermalize the system at arbitrarily low temperatures. For a fixed choice of move class, the objective function might have a multitude of suboptimal solutions, each acting as a dynamical trap from which the system does not escape on time scales of interest. At first glance, it appears that this trapping problem can be circumvented by choosing a wide move class—in the extreme limit, the neighbor list could include the whole state space. Such choice would remove the very existence of local minima, at the price of turning simulated annealing into random search, which is a poor optimization strategy. In the language of physics, one would have replaced energetic barriers with entropic ones.

An energetic barrier leads to a high rejection rate of attempted moves out of a local minimum energy configuration due to the Boltzmann factor in the transition probabilities. The effect of an entropic barrier is slightly more subtle. There might be a large number of dynamically available paths. However, an overwhelming majority among these does not lead to the desired solution, which plays the role of the proverbial needle in a haystack. For a fixed move class, the relative importance of entropic versus energetic barriers is determined by the temperature parameter, with energetic barriers dominating the low-temperature behavior.

To design move classes that are a good trade-off between energetic and entropic barriers, a degree of understanding of the structure and geometry of configuration space seems highly desirable. At a more academic level, one could also ask the reverse question: What is the reason for the success of simulated annealing as an optimization algorithm, in the face of the above-mentioned potential pitfalls? Are there general emerging features of complex configuration spaces of interesting problems, which can be used to (at least partly) classify the geometry and the corresponding relaxation behavior?

In this chapter, we first discuss in general terms how this theme can be approached. Second, we introduce a number of concepts related to coarse-graining (of state space and the dynamics), which is often the appropriate tool for understanding relaxation in multiminima systems. Third, we provide a short description of particular cases that have been studied, together with pointers to the relevant literature.

15.1 The Coarse Structure of the Landscape

We are interested in hard optimization problems possessing a very large set of highly interconnected states. The set is so large that any sampling performed in a realistic amount of time can only include a vanishingly small fraction of them. Many problems of practical interest fall into this category.

To set the stage, let us first think of a search of t_m steps carried out by a Metropolis random walker, at a constant, perhaps even quite high, temperature. The search is repeated n times, each time with a new, randomly chosen initial condition. Intuitively, we expect that if t_m is sufficiently large, the statistical properties of the results obtained in different repetitions should not vary a great deal among the set of trials. If this indeed is the case, the state space must consist of statistically similar subsets, each of which can be sampled in t_m steps.

For illustration purposes, let us consider the toy example depicted in Fig. 15.1, where the state space is the real axis and the energy function E is sinusoidal with an added tiny perturbation making the minima inequivalent; e.g., we write $E(x) = b/2 \sin(x/L) + \sum_i \varepsilon_i(x)$, where b is a positive constant and where $\varepsilon_i \ll 1$ is nonzero only in a neighborhood of the ith minimum of the sine function. In this case, t_m is the time it typically takes a random walker to explore just one basin of the objective function; t_m may depend on the temperature or, if annealing is performed, on the schedule used or any other details. Whichever search strategy is used to find the minima, sampling beyond t_m will yield a random sequence of values, say, $-b/2+\delta_i$, with a narrow distribution of δ's, peaked at zero. The structure of the problem—in this case the sinusoidal shape—can cleverly be utilized to find the important part of the energy minimum, which is $-b/2$. Optimizing beyond this level amounts to sifting through a series of n independent random numbers, an undertaking whose rate of success (the rate at which better solutions are found) decreases as n^{-1}! The point here is that if there is no structure in the landscape beyond a correlation time t_m, no particularly clever way exists for carrying out the optimization beyond that time.

The concept of correlation time for a random walker exploring the complex landscape of optimization problems was quantitatively analyzed by Stadler [Sta92]. Stadler considered the autocorrelation of a time series E_t of energy values that are sampled by a nearest-neighbor random walk on the state space of several combinatorial problems: graph bipartitioning, graph matching, and traveling salesman. The random walk is performed at infinite temperature, i.e., with all moves accepted.

Stadler found that the autocorrelation function has an exponential form, with a scale parameter, the correlation time, growing linearly with the size of the problem. This scale parameter corresponds to our t_m. His interesting result has some bearing on the issue of parallel versus sequential exploration of the landscape more fully discussed in Chapter 8. Simply stated, the results suggest that it would be better to place an independent walker in each of the correlated volumes rather than using the same amount of computer time to do one correspondingly longer walk. Each walker would by a clever strategy find her best minimum, and the best of these can then be selected as the solution of the problem.

Summarizing, we surmise that a smallest time scale t_m exists, such that for all times $t > t_m$ subsets of state space sampled in t steps by walks starting from different, randomly

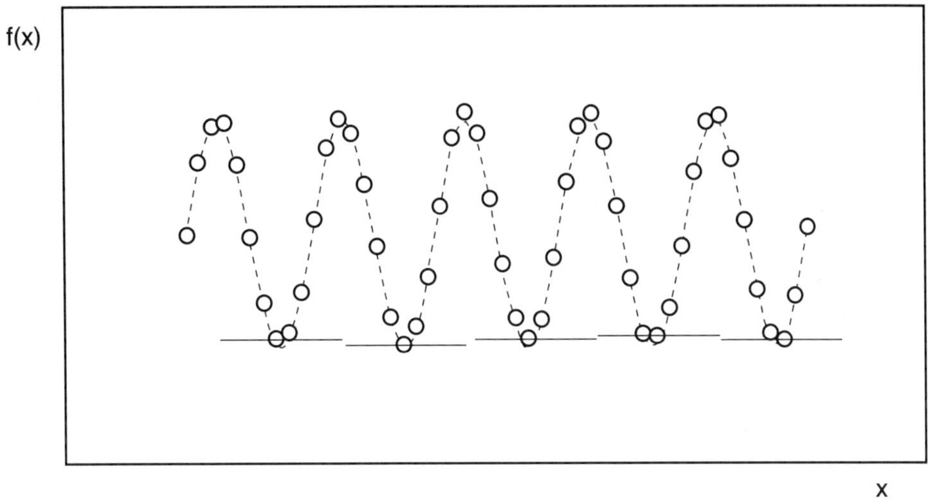

Figure 15.1. *A sinusoidal energy function with small random perturbations.*

chosen initial conditions, are statistically equivalent. If all the states belonging to one of these subsets are lumped together, the resulting coarse-grained landscape has no structure left. Therefore, simulated annealing or any other optimization technique relying on local information will not work well beyond t_m. For sampling times smaller than t_m the structure of the landscape yields information that can be useful in the search. This is the lesson of the brick wall effect described in Chapter 8.

15.2 Exploring the State Space Structure: Tools and Concepts

As just discussed, the statistical properties of a time series E_t of energy values sampled by a random walker at $T = \infty$ already give some useful information on the coarse structure of state space. Repeating the analysis for different values of the temperature T would sample different portions of the landscape and presumably would yield different values of the correlation time for each T.

While Gehring [Geh97] extended this approach to finite but large temperatures, an investigation of the low-temperature behavior along these lines has not yet been performed. Nevertheless, the number of a priori possible scenarios is quite limited. Decreasing the temperature makes all transition probabilities per unit time smaller, or leaves them unchanged. Hence, the system must linger for a longer time in any particular region. The average correlation time of the series must therefore increase with decreasing temperature. However, it could either remain finite at all temperatures or diverge at some critical temperature T_g. This latter case implies the possibility of a qualitative change of relaxation behavior at low T_g. We argue below that many hard combinatorial systems have a glassy behavior below a well-defined critical temperature, where correlation functions typically decay algebraically rather than exponentially. (See also the discussion in section 6.6.)

Since at low temperature the transition rates out of a local energy minimum configuration are small, a set of states, called a *trap*, is repeatedly visited. After a while the trajectory exits the trap and new regions of state space are explored. The average exit time may correspond to a very large number of elementary moves. If such *separation of time scales* between the time scale of the elementary move and the exit time applies, the system is able to reach a state of so-called local equilibrium.

In a locally equilibrated system, the probability $p(x, t)$ of finding the system in state x at time t is in general time dependent, unlike true equilibrium. However, for any two states x and x' within the trap, the ratio of probabilities $p(x, t)/p(x', t)$ is very close to the Boltzmann ratio $\exp[(E(x') - E(x))/T]$ (see Example 6.9), similar to an equilibrium situation. In other words, all probabilities $p(x, t)$ for x's in the trap share the same, so far arbitrary, time dependence, which factors out when calculating ratios.

Within some error bound, the net flow of probability in or out of the trap only changes the probability distribution inside it through the time dependence of a normalization factor. This makes it possible to introduce an enormous simplification: the internal degrees of freedom of the trap are neglected, and its states are lumped together into just one effective state. Applying the lumping procedure to all the traps of the system state space yields a new set of states that together form a *free energy landscape*. This coarse-graining approach is very well known in chemical physics, dating back to at least Kramers [Kra40]. A recent application to complex systems, where the lumping is explicitly carried out, can be found in [SS94, SSA95].

The following questions immediately arise: Will the dynamics on the new coarse-grained state space—the free energy landscape—be Markovian? And if so, what are the hopping rates? Strictly speaking, the answer to the first question is negative [BR58]. Nonetheless, we would claim a Markovian approximation is very useful, despite some limitations. A full discussion of these issues will take us too far afield, so instead we use intuitive arguments on the simple example already introduced, where the state space is the real axis and the energy function, $E(x)$, is sinusoidal. The example serves us well for illustration purposes, but the reader should be warned that the one-dimensional character of the problem makes the results partially misleading, since strong entropic effects are present in high-dimensional problems. Some issues related to these effects are discussed in the Appendix to this chapter.

As discussed in Chapter 2, continuous and discrete versions of optimization problems often have very similar properties. Here we find it convenient to start with a continuous description in time and space. The coarse-grained procedure we are about to outline leads to hopping dynamics on a discrete space: the free-energy landscape of the problem.

The free-energy landscape of the coarse-grained problem consists of a discrete set of points, each representing one trap, namely, the interval between two adjacent maxima. Within the ith interval the probability distribution in the previously mentioned local equilibrium approximation has the form

$$P(x, t) = f_i \exp[-E(x)/kT] \tag{15.1}$$

with f_i varying slowly on the microscopic time scale of the problem. The slow variation applies if the temperature is so low that the escape rates out of well i and into the adjacent wells $i \pm 1$,

$$k_{i+1,i} = k_{i-1,i} = \exp(-b/kT), \tag{15.2}$$

are much smaller than one. We recall that the parameter b is the difference between the energy of the bottom of a local minimum and the energy of the rim leading to a neighboring basin.

The coarse-grained Markovian approximation of the dynamics simply consists of a random walk on the free-energy landscape with transition rates $k_{i\pm1,i}$. In a slightly less trivial example, the right and left barriers could be different, resulting in a biased random walk. In our case the random walk can also be treated as a new diffusion process with a dressed diffusion coefficient

$$D' = D \exp(-b/kT). \tag{15.3}$$

This so-called activated diffusion is widespread in physical systems and occurs, e.g., in diffusion of impurities in the periodic potential of a crystal structure [McQ00].

The subdivision of state space into different disjoint traps relies on drawing a somewhat arbitrary line separating nearby configurations of the original problem: what appears at the lumped level as a hop between adjacent traps is in fact just a tiny change of position, which happens near a trap boundary. The system will therefore tend to repeatedly cross the same boundary, rather than moving on to an altogether different trap. This shows an effect we had anticipated: the coarse-grained dynamics (of our toy example) cannot be exactly Markovian, as correlations are present on microscopic time scales.

To circumvent the problem one can perform an additional coarse graining in time, by choosing $\tau_A = \exp(b/kT)$ as the smallest time scale of the coarse-grained dynamics. At times that are multiples of τ_A, the system is almost certainly close to one of the minima and has lost the memory of which minimum was visited in the previous time step, by virtue of the fluctuations performed at the bottom of the well. Therefore, on these coarser time scales the dynamics again appears Markovian.

In a high-dimensional—and for applications far more relevant—version of the same problem, the region separating adjacent traps is not the maximum of the energy function as in one dimension but is rather a saddle point. If the energy increases along most paths emanating from the saddle point, the surrounding region may possess a high degree of metastability. In this case, when the system moves through the saddle on its way from one low-energy region to another, the memory of the origin of the trajectory is partially erased by thermalization in the saddle region.

In general, the problem at hand must be carefully analyzed for one to decide if separation of time scales and a Markovian description of the coarse-grained dynamics may apply. However, at least in physical systems, metastability and separation of time scales appear to be the rule rather than the exception.

A peculiarity of one-dimensional systems is hidden in the form of the rates. In general, according to Kramers' classic argument, the escape rate out of a trap is proportional to the value of the quasi-equilibrium probability taken at the edge of the metastable region. Thus the rate from trap j to trap i has the form

$$k_{i,j} = \exp(-b_{i,j}/kT) \cdot \mathcal{D}_{i,j}, \tag{15.4}$$

where $b_{i,j}$ is the energy barrier separating the two traps, and $\mathcal{D}_{i,j}$—the degeneracy—is the number of states or volume in the saddle region connecting i and j. This prefactor, which in the example is just 1, can in some interesting cases play a very important role in the system dynamics.

Summarizing this section, we have qualitatively described a coarse-graining technique by which the fast relaxing degrees of freedom corresponding to relaxation within a trap are separated from the slow degrees of freedom pertaining to the flow of probability from one trap to the other. The information passed on from the fine-grained to the coarse-grained level of description concerns the geometric properties of the traps: the local density of states (number of states at the trap edge) and the height of the (activation) barrier separating the traps.

A direct demonstration of the applicability of the above ideas for problems beyond toy examples is not a simple task, and most of the (strong) evidence for the scenario is actually indirect. There have been, however, several attempts to characterize the free-energy landscape of selected examples beyond the correlation studies already cited: molecular clusters [Ber93, BBK+96], spin glasses, and combinatorial problems have been investigated. In the next section, we examine some of these approaches.

15.3 The Structure of a Basin

Since at low temperatures the system is usually confined to neighborhoods of local minima, local geometrical properties within these neighborhoods are the key to understanding important aspects of the relaxation behavior. To numerically investigate these properties, one needs an operational definition of neighborhood in state space.

To this end, consider a low-energy minimum or reference state ω_0. By shifting the energy, $E(\omega_0)$ can be set equal to zero. For some energy value $L \geq 0$ and a fixed temperature T, the system will typically not reach states with energy higher than L on time scales smaller than the Arrhenius time scale $t_A = \exp(L/T)$. It is therefore useful to define an L neighborhood of ω_0 as the set of states that can be reached through a set of moves that exclude states of energy larger or equal to a *lid* L.

Such a basin is a subset of state space, with a size or volume $\mathcal{V}(L)$, which grows as function of L. Banning the unlikely possibility of dynamically connected degenerate states at zero energy, we have $\mathcal{V}(L = 0) = 1$, i.e., the pocket contains only the reference state when the lid L is zero. By continuity, $\mathcal{V}(L)$ will have a reasonable size for small enough L, and an exhaustive enumeration of all its states is a tractable task, at least for moderate values of L.

Besides $\mathcal{V}(L)$ it is also of interest to investigate the local density of states $\mathcal{E}_L(E)$, $E \leq L$, which is the number of states per unit energy at energy E. Finally, one can study the relaxation dynamics within the pocket in full detail by keeping track of the connectivity of the system, i.e., by maintaining for each state a list of all its neighbors, together with their energies. With this information, the transition matrix of the Markov chain describing the relaxation process in the basin can be constructed and the time-dependent probability distribution calculated for any choice of initial conditions.

Before entering a discussion of complex landscape geometry, let us briefly consider how our definition works for a couple of rather simple cases: the sinusoidal potential and a random potential. In the former case, where, as in any continuous system, the number of states is uncountable, we just discretize the state space by specifying positions with a finite accuracy.

Consider first the sinusoidal example. If we pick any local minimum as reference state and count all the states below the lid, we find that $\mathcal{V}(L)$ starts increasing in a rather smooth fashion. (The details will depend on the mesh size of the grid chosen.) This behavior persists up to the lid value $L = b$, above which the whole state space suddenly becomes available: $\mathcal{V}$ jumps then to a very large value, which is the total number of states in the system.

Another simple system is a random energy landscape, where each state has a random energy drawn from some distribution. Again, we pick a local energy minimum and watch $\mathcal{V}(L)$ grow as L grows. At some L value there will be a percolation transition, meaning that a nonzero fraction of the whole system becomes available [Cam85, SA94]. At that particular lid value $\mathcal{V}$ will again jump to a very large (possibly infinite) value, which is similar in a way to our sinusoidal example.

15.4 Examples

We now turn to the complex problems that have already been analyzed according to the lid algorithm—traveling salesman problem [SSSA93], spin glasses [SS94, Sib98], and glasses [SS98]. The available volume $\mathcal{V}(L)$ grows in an approximately exponential fashion, as shown in Fig. 15.2. Jumps (i.e., discontinuities on the scale of $\mathcal{V}$) do occur as new side valleys join the volume already discovered. Unlike the previous gedanken examples, the percent change in $\mathcal{V}$ due to the discontinuities is quite small, so that the function has an overall smooth appearance, which is close to a straight line in a semilogarithmic plot.

The local density of states $\mathcal{E}_L(E)$ also grows in a close-to-exponential fashion. This means that new states emerging at each increment of L predominantly have energies close to L. In other words, many of the side valleys are rather shallow.

To understand the origin of this close-to-exponential behavior, it is convenient to consider first the extreme limit, where a unique path connects any two states. In this case, the topology of state space is, by definition, that of a tree, and including new states in the pocket by increasing the lid is akin to a branching process, which generically has an exponential growth law.

This idealized situation cannot exist in high-dimensional state spaces, where in general very many different paths connect pairs of states. However, if the overwhelming majority of these paths is energetically forbidden, as would typically be the case for Metropolis walks at low temperatures, loops in state space will be relatively rare, and the no-loop idealization can serve as a useful approximation. Based on these considerations, and encouraged by the fact that the very different examples analyzed so far all behave in a roughly similar way, we conjecture that close-to-exponential-growth laws will be a typical feature of highly frustrated optimization problems.

Assuming the validity of the above scenario, we now explore its consequences for the relaxation behavior, as described by the Kramers approach. Readers interested in a more detailed analysis and more examples are referred to [SSSA93, SS94, SS00]. Here we simply note that the local equilibrium distribution within the basin has the form

$$P_{(x,t)} \propto \gamma(t)\mathcal{E}_L(E)\exp(-E/T), \tag{15.5}$$

where $\gamma(t)$ is the time-dependent probability of being in the trap. If the local density of states has the exponential form $\mathcal{E}_L(E) \propto \exp(E/E_0)$, the Kramers approximation will work only for $T < E_0$.

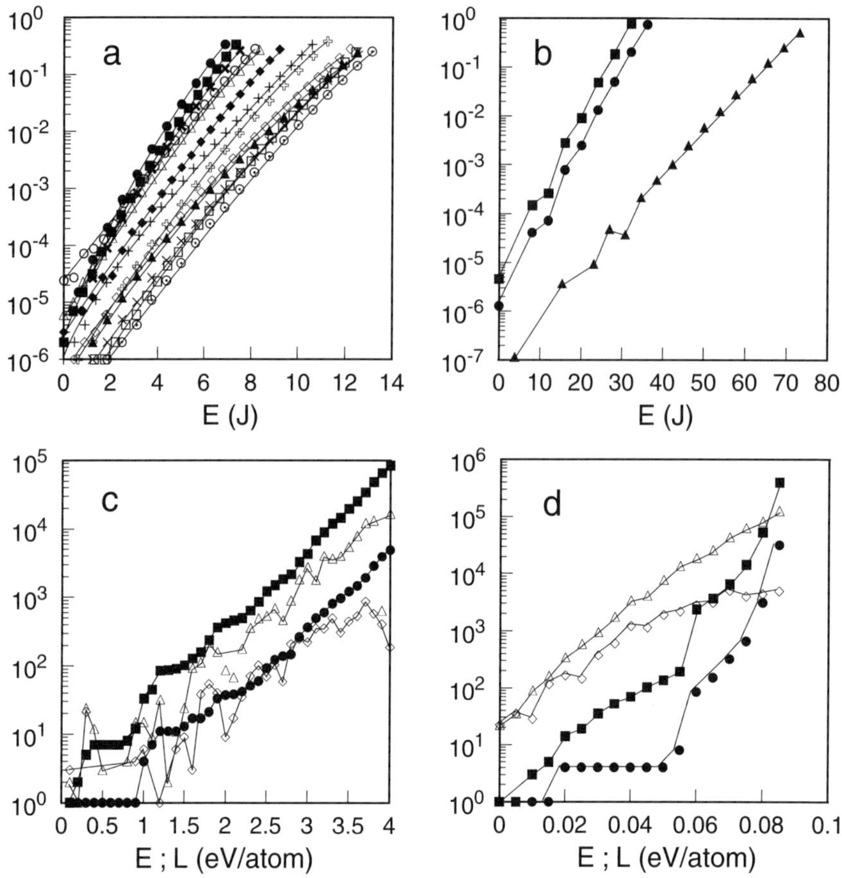

Figure 15.2. *The exponential growth of local densities of states. This figure, taken from* [SvdPS99], *illustrates the geometry of the landscape close to energy minima in different systems.* (a) *The local densities of states* $\mathcal{D}(E, L_{\max})$ *for* 12 *realizations of a spin-glass model on a* $5 \times 5 \times 5$ *lattice.* (b) *The same quantity for the ferromagnetic Ising model. The circles are for an* 8×8 *lattice, the squares for a* 7×7 *lattice, and the triangles for a* $4 \times 4 \times 4$ *lattice. In all cases the ordinates are normalized by the total number of states found and the abscissa is the total energy in units of* J. *The exponential growth parameters are close to the transition temperatures of the corresponding systems.* (c) *and* (d) *show data for two-dimensional network models of a glass and of a polymer system, respectively. Here, the abscissa is either the energy* E *or the lid* L, *both expressed in (eV/atom). The curves shown are: the available volume* $\mathcal{V}(L)$ *(squares) and the number of minima* $\mathcal{M}(L)$ *(circles) as a function of the lid; the local density of states* $\mathcal{D}(E, L_{\max})$ *(triangles) and the local density of minima* $\mathcal{DM}(E, L_{\max})$ *(diamonds) as a function of the energy.*

Indeed, for $T > E_0$ the local equilibrium distribution would be biased toward high-energy values rather than low ones, which would violate the Kramers assumption that the probability flux out of the trap is small. In physical terms we say that the trap loses its metastability at $T = E_0$. This happens regardless of the depth of the trap. By way of contrast, a trap with, for instance, a power-law density of states has a trapping effect that depends on the depth of the trap and decreases continuously with the temperature rather than abruptly.

To gain a more detailed understanding of the relaxation dynamics, it is possible to study mesoscopic models, e.g., random walks on tree structures, which have the exponential growth laws built in as a modeling assumption. Such studies have been undertaken by a number of authors [OS85, GWH85, SH89]. A recent comprehensive numerical and analytical study can be found in Uhlig [UHS95]. The essence of these studies is that the relaxation is algebraically slow below the transition temperature. In addition, the relaxation features temperature-dependent decay exponents and rather complicated memory effects. From the point of view of optimization, such behavior is highly undesirable but probably unavoidable, although one can try to postpone its onset by choosing a move class with a lower T_g.

15.A Appendix: Entropic Barriers

This short Appendix is devoted to the exploration of some exact mathematical results bearing on the issue of so-called entropic barriers. In physical parlance, the entropic barrier is the logarithm of the relaxation time multiplied by the temperature. In everyday language it can be thought of as describing a needle in a haystack or a bottleneck effect: all paths in the graph are equally probable, but very few lead from one given set of states to another. Here, we consider state spaces on which the energy is constant, and thus the only structure on these spaces is the graph structure given by the move class. We then ask what the largest the relaxation time can be on such graphs as a function of the size of the graph and its connectivity.

This question bears on the issue of thermal relaxation and hence annealing in more realistic multiminima landscapes for the following reason: Throughout this chapter, and indeed throughout most of this book, we rely heavily on a local thermal equilibrium approximation within traps or valleys in the energy landscape. This means that the Markov process describing the dynamics must have sufficient time to relax within each valley before any sizable outflow of probability occurs through the rim of the valley. In other words, the relaxation time of the internal dynamics must be shorter than the characteristic escape time given by the Kramers formula. Although strong empirical evidence supports the local equilibrium assumption, theoretical probes of this question might entice the more mathematically oriented reader. Readers more interested in application can safely skip this Appendix.

For valleys with an exponential local density of states, (15.4) gives us an escape time τ_{esc} (the inverse of the escape rate), which either grows exponentially ($T < E_0$) or decays exponentially ($T > E_0$) with the energy (gauged from the bottom of the valley) and hence also with the the number N of configurations that our valley contains. As previously discussed, only the low T regime is of interest for the analysis, and in this regime $\tau_{\text{esc}} \propto N^{E_0/T-1}$. Considering now that the low-lying states are those playing the most important role for the internal relaxation, it seems reasonable to compare with the N dependence of the relaxation time ε for a relatively flat valley with the exponential N dependence of τ_{esc}.

Neglecting logarithmic corrections, the mathematical analysis presented in the sequel shows that $\varepsilon \propto N^x$, where x is close to one for sparse graphs. Hence, for large enough valleys and for sufficiently low T we can expect $\tau_{\text{esc}} > \varepsilon$. This explains why the local equilibrium approximation can be confidently applied for sufficiently low T. This regime is bounded not by E_0 but by a lower temperature $T' = E_0/(1 + x)$. In the specific context of a random walk on a hierarchical structure [Sib87], this conjecture can be shown more precisely and T' can be given an interpretation: For $T < T'$ the random walk is recurrent (and thus equilibrium-like), while for $T > T'$ the random walk is transient.

The connectivity issue may seem irrelevant at first sight. After all, when dealing with combinatorial problems by simulated annealing techniques one always ends up with regular graphs, where all states have the same connectivity. Not so, however: As already hinted, at low T most states of the graph are effectively unavailable on reasonable time scales, because their energy is too high. What remains of state space when the unavailable states are removed is a graph with fewer states and with a lot of variation in connectivity.

15.A.1 The Master Equation

First, it is convenient to generalize the relaxation equation to the case in which time is continuous rather than discrete. This amounts to considering a Markov process instead of a Markov chain [Kam92, Fel66b]. The shift is most easily achieved by first expressing the transition probability of the chain as[1]

$$M = \exp(W), \tag{15.A.1}$$

whence the time dependence of the probability distribution becomes

$$p(t) = \exp(tW) \cdot p(0). \tag{15.A.2}$$

We now let t be an arbitrary real positive quantity, differentiate the above equation with respect to t, and find

$$\frac{dp}{dt} = e^{tW} W p(0) \tag{15.A.3}$$

$$= W e^{tW} p(0) \tag{15.A.4}$$

$$= W p(t). \tag{15.A.5}$$

In the physical literature [Kam92], (15.A.5) is called the *master equation* and plays an important role, because the entries of W can often be calculated from basic physical principles, e.g., quantum mechanics.

Expanding the exponential in (15.A.2) to first order in t, we easily see that the quantities tW_{ij} are the probabilities that a transition occurs from state j to i during a very short time interval t. The entries in W are therefore called transition probability densities. We also note that the column sums of W must vanish and that the eigenvalues of W are given as $\lambda'_i = \ln \lambda_i$, where λ_i are the eigenvalues of M. Since $0 \le \lambda_i \le 1$, it follows that $-\infty < \lambda'_i \le 0$, with the equilibrium eigenvalue being $\lambda'_1 = 0$. The relaxation time is given according to (6.22) by $\varepsilon = -1/\lambda'_2$.

15.A.2 Random Walks on Flat Landscapes

Consider a nearest-neighbor random walk along the edges of an arbitrary connected graph G, with all transition probabilities equal. In this case, both M and W are symmetric matrices. In particular, modulo an arbitrary common multiplicative constant, for $i \neq j$, $W_{ij} = 1$ if i and j are connected by an edge, and zero otherwise. Due to the zero column sum rule, the diagonal elements are given by $W_{ii} = -n(i)$, where $n(i)$ is the number of edges emanating from node i. We note that since each column sum of W vanishes, $\mathbf{e} = (1, 1, \ldots, 1)$ is a left eigenvector with eigenvalue zero. Due to the symmetry of W, its transpose $\mathbf{e}^\dagger$ is a right eigenvector with the same zero eigenvalue. Apart from a normalization constant, the uniform distribution $\mathbf{e}^\dagger$ therefore describes the stationary solution of the problem. The time scale $\varepsilon = -1/\lambda'_2$ on which equilibrium is established depends on the size and connectivity of the set of states viewed here as a graph.

[1] The exponential function for matrix arguments is defined in terms of the Taylor series $\exp(W) = I + W + W^2/2! + W^3/3! + \cdots$, where I is the identity matrix.

Before looking at the general bounds described in the next section, consider the following two simple analytically soluble cases: the nearest neighbor walk on a set of integers $0, 1, \dots, N$ and the walk on a fully connected graph.

EXAMPLE 15.1. In this example we consider the random walk on the integers $0, 1, \dots, N$ with only nearest-neighbor connectivity and with reflecting boundaries. The master equation is

$$\frac{dp_i}{dt} = \frac{1}{2}(-2p_i + p_{1+1} + p_{i-1}), \qquad 1 \le i \le N-1. \tag{15.A.6}$$

The endpoints $i = 0$ and $i = N$ have different equations because of the presence of the reflecting boundaries. However, we can get around this difficulty, by considering the *even* solutions of an allied problem, defined by the very same equations, but on the set $-(N-1) - (N-2), \dots, 0, \dots, N$, and with periodic boundary conditions, i.e., where the points $-N$ and N are identified. The even eigenvectors of this latter problem are, for $k = 0, 1, \dots, N-1$,

$$\phi_k(j) = \cos(\pi jk/N), \tag{15.A.7}$$

and the corresponding eigenvalues are

$$\lambda'_{k+1} = \cos(\pi k/N) - 1, \tag{15.A.8}$$

where we have respected the convention that the first eigenvalue is zero. In particular, the relaxation eigenvalue is $\cos(\pi/N) - 1$, which in the limit $N \to \infty$ can be approximated by $\frac{1}{2}(\pi/N)^2$.

EXAMPLE 15.2. The master equation for the fully connected graph with N nodes has the very simple form

$$\frac{dp_i}{dt} = \frac{1}{N}(1 - Np_i) \ \forall i. \tag{15.A.9}$$

The prefactor $1/N$ ensures that the average number of transitions per unit of time is of order one. Note that the system of equations is already diagonal. As all p_i's obey the same equation, only two different eigenvalues can be present. The first one is zero and the second is the highly degenerate relaxation eigenvalue. Writing $p_i(t) = 1+\delta_i(t)$, we find $\delta_i(t) = \delta_i(t=0)\exp(-t)$. Hence, in these units, the relaxation time is $\varepsilon = 1$, independent of the size of the graph.

15.A.3 Bounds on Relaxation Times for General Graphs

A general bound for the value of $|\ \lambda'_2\ |$ for a walk on a general graph G was derived by Thomas and Yin [TY86]. To discuss their main result we first mention a standard (and intuitively clear) theorem, also described in [TY86]: Removing an edge from G makes the relaxation time larger, i.e., pushes the relaxation eigenvalue λ'_2 closer to zero.

To obtain a bound, we can then keep removing edges until we obtain a spanning tree. Going beyond that, one would obtain a disconnected graph and change the spectrum qualitatively, as the zero eigenvalue would become degenerate. We also need a couple of additional notations. We let g_τ be a tree graph (i.e., a graph containing no closed paths), which spans the original graph G. This means that g_τ and G have the same nodes, albeit in general not the same edges. If one removes a node i from g_τ along with the $n(i)$ edges emanating from it, one obtains a set of $n(i)$ disconnected branches. The number of nodes in each of these branches is denoted by $\mid b_i(j, g_\tau) \mid$, $j = 1, 2, \ldots, n(i)$. Finally, we let $d_i(j, g_\tau)$ be the diameter of the branch, which is the length of the longest path it contains.

The bound proved by Thomas and Yin is

$$\mid \lambda'_2 \mid \geq \sup_{i \in g_\tau} \inf_j \frac{1}{(1 + d_i(j, g_\tau)) \mid b_i(j, g_\tau) \mid}. \tag{15.A.10}$$

The following examples serve as simple illustrations:

1. The previously considered case where G is set of integers $0, 1, \ldots, N$. As the line is its own spanning tree, to find the bound we can remove the central j node, thereby obtaining two equal branches. Assuming for simplicity that N is odd, we have $|b_i(j, g_\tau)| = d_i(j, g_\tau) = (N - 1)/2$. Hence $\mid \lambda'_2 \mid \geq 4/N^2$, or $\varepsilon \leq N^2/4$, which has the same N dependence as the exact result.

2. G is a Cayley tree, with coordination number z. Such a tree can be drawn starting from a central point at level 0 to which z level 1 points are connected. At level j there are z^j points. Adding M layers we have, all in all, $N = (z^{M+1} - 1)/(z - 1)$ nodes. All the internal nodes have z neighbors, while the last tier only has $z - 1$ neighbors. This is a graph with a sparse connectivity, which is of course its own spanning tree. Removing the central nodes leaves us with z subtrees, each with diameter $M - 1$ and size $(z^M - 1)/(z - 1)$. For large M we get $\mid \lambda'_2 \mid \geq z \ln(z)/(N \ln(N))$. Basically, the relaxation time diverges as $N \ln N$, which, for large N values is a close to a linear dependence.

3. G is a fully connected graph with N nodes. In this case, a spanning tree g_τ can be drawn by choosing any node as a central node, from which $N - 1$ edges emanate to each of the remaining nodes as spokes from a hub. If we pick the hub as the node j to remove from g_τ we obtain $N - 1$ disconnected branches, each of length 1. Hence $d_i(j, g_\tau) = 1$, $\mid \lambda'_2 \mid \geq 1/2$, and $\varepsilon \leq 2$. We see that in a fully connected graph, the relaxation time is of order one, independent of the size of the graph, as in the exact result.

4. G consists of two blobs, each of size N and fully connected, plus an additional central node. The two blobs communicate via two edges through the central node. If we remove this one node, we see by the same argument as before that $d_i(j, g_\tau) = 1$, while $\mid b_i(j, g_\tau) \mid = N$. Hence, $\mid \lambda'_2 \mid \geq 1/(2N)$ and $\varepsilon \leq 2N$.

The conclusion we can draw from from the above is that on large and sparsely connected graphs, the relaxation time can be expected to diverge in a close-to-linear fashion with the size of the system's state space.

Chapter 16

What Makes Annealing Tick?

Admittedly, this question is not quite well defined. Nevertheless before attempting an answer, let us limit the scope of the claim it implies by re-emphasizing that the success of an optimization algorithm strongly depends on the amount of knowledge and effort that goes into designing it. Simulated annealing is a simple heuristic that is generally applicable to optimization problems with only a minimum of prior knowledge and programming effort.

Nevertheless, simulated annealing usually does much better than a greedy algorithm, which quickly gets caught in a high-lying local minimum. Some features that contribute to this are as follows:

- The stationary distribution at any T is uniform in energy, i.e., all states at a given energy are equally likely.
- The sequence of stationary distributions are as typical as possible, consistent with a given $\langle E \rangle$.
- Annealing allows uphill moves, making it possible to leave the basin of attraction of a minimum after entering.

The first two features concerned much of the discussion in Chapter 5. Before we try to combine these facts with the arguments and structural information presented in Chapter 15, we pause to examine the extent to which the third feature can help.

16.1 The Dynamics of Draining a Basin

Let us now try to understand more quantitatively the way in which a system subject to a cooling schedule $T(t)$ is eventually trapped in a local minimum. To do this we need a concept developed in Chapter 15—the idea of local thermal equilibrium in a basin.

Assume that the basin has a depth ΔE. This means that to leave the basin, the system has to visit a state whose energy is at least ΔE. We also let $\mathcal{D}$ be the local density of states as a function of the energy. This means that $\mathcal{D}(x)dx$ is the number of states in the basin having energies in the interval $(x, x + dx)$.

The Kramers argument (see section 6.5.2) gives us a rate of flow r out of the trap equal to

$$r(t) \propto \exp\left(-\frac{\Delta E}{T(t)}\right)\mathcal{D}(\Delta E). \tag{16.1}$$

The formula is a good approximation only as long as the rate of flow out of the trap is small, i.e., if $r(t) \ll 1$.

Assume now that the system is initially in the trap and that a local thermal equilibrium situation has been established and can be maintained throughout the slow cooling process. The probability of the system still being trapped at time t, $P(t)$, fulfills the differential equation $\frac{dP(t)}{dt} = -r(t)P(t)$, which has the solution

$$-\ln\frac{P(t)}{P(0)} = \int_0^t r(t')dt'. \tag{16.2}$$

If (and only if) the integral on the right-hand side diverges as $t \to \infty$, then $\lim_{t\to\infty} P(t) = 0$. This means that the trap will eventually be left with probability one.

As the temperature is lowered, $r(t)$ must go to zero. The fastest possible decay still leading to a divergent integral [Rud76] is in practice[1]

$$r(t) = 1/(t+1), \qquad t > 0. \tag{16.3}$$

Accordingly, the fastest schedule guaranteeing this result is

$$T(t) = C/\ln(t+1), \qquad C > \Delta E. \tag{16.4}$$

In a finite system all barriers are finite. We can now (in principle) choose C larger than the largest barrier in the system to make sure that all traps will eventually be escaped. This leads to the Geman and Geman result [GG84] described in Chapter 13, stating that a logarithmic cooling schedule eventually finds the ground state with probability one.

The purpose of presenting this short heuristic argument for the Geman and Geman schedule is mainly to emphasize that for any realistic schedule the system will eventually be trapped in a local minimum, which probably will not be the ground state. In this situation the empirical success of simulated annealing appears rather puzzling. It must hinge on additional properties of the energy landscape that are shared by a large number of different systems. In the next section we discuss what these properties might be.

16.2 Putting It Together

We now combine the characteristics listed at the beginning of this chapter with the following structural conjectures:

- The local density of states in a basin is approximately exponential in the energy.

[1] We are here neglecting the functional forms: $r(t) = 1/(t\ln(t))$, $r(t) = 1/(t\ln(\ln(t)))$, $r(t) = 1/(t\ln(\ldots\ln(\ln(t))\ldots)$ etc., since these iterated logarithms will be nearly constant on the time scales accessible to a numerical simulation.

- Deep basins have large rims.
- The state space of combinatorial problems consists of many statistically equivalent regions.

The first of these characteristics was discussed in Chapter 15 and has been seen empirically using the lid method for several systems, as shown in Fig. 15.2 and surrounding discussion. It is our belief that it is a rather general feature of complex landscapes. Recall that an exponential local density of states implies that above a critical temperature (equal to the exponential growth constant E_0 in $\mathcal{D}(E) \sim \exp(E/E_0)$), the probability distribution over the states in the trap is strongly biased in favor of high-energy states close to the rim. Below this temperature the distribution has most of its mass close to the bottom.

The consequences of an exponentially growing local density of states for annealing were analyzed by Schön [Sch97]. We can assume with this author that once a walker enters a trap with a local transition temperature higher than the current temperature—as set by the annealing schedule—the walker is trapped and never leaves again.

The second characteristic is a corollary of the exponential growth. It may be proved as a theorem for smooth functions on $\mathbb{R}^n$ [ZS92] and directly for a number of examples, including graph bipartitioning [Ven92]. This characteristic explains what happens when different basins compete for the same walker. In [Sch97] it is assumed that the system is in a state of approximate thermodynamic equilibrium at the rim. Under these conditions, as the temperature is decreased the deepest trap is typically the one in which the walker gets caught most often, since it has the largest basin of attraction. Thus an explanation for the success of simulated annealing, at least for systems where the local density of states is approximately exponential, is that

Simulated annealing results in preferential trapping into deep basins.

The third characteristic ensures that although we can sample the landscape of the problem only very sparsely, we can still find a good solution to our problem within any region of the landscape. This of course applies equally to any search technique.

16.3 Conclusions

We have presented some heuristic arguments for the surprising success of simulated annealing, all relying on structural properties of the landscape that have been observed for a number of examples but that remain to be proved for the general case. As yet, there does not exist a good classification scheme for complex optimization problems based on the structural properties of their landscapes. However, some work has been done, which can be seen as a start in this direction. We believe that this work will ultimately bring the subject from the realm of empiricism to the realm of provably optimal algorithms.

Part V

Resources

Chapter 17
Supplementary Materials

As a closer study of the references of this book shows, complex optimization and the structure and classification of complex energy landscapes remain active areas of research. Consequently, many resources such as software and additional references are available on the Internet. Interested readers will find in this chapter a commented list of resources.

Additional material more directly related to the contents of this book can be found at `http://www.frostconcepts.com/`.

Specific questions regarding errata and updates can be addressed to `salamon@math.sdsu.edu`.

17.1 Software

17.1.1 Simulated Annealing from the Web

A simple search on a web search engine will result in numerous hits for "simulated annealing software".[1] For example, a search using `www.google.com` resulted in about 21,000 hits. On closer inspection, the reader will find that the vast majority of these hits are university projects implementing bare-bones annealing for single- or multiprocessor systems. Because bare-bones annealing is embarrassingly parallel, it is a common target for developers of parallel scientific software libraries.

Of the university development efforts, Dyke Stiles and a dedicated group at Utah State University maintain solid implementations of single- and multiprocessor bare-bones simulated annealing software along with a variety of other optimization heuristics:

`http://www.engineering.usu.edu/ece/research/rtpc/projects/comb/`.

Lester Ingber has long been a source for robust simulated annealing software and reference materials—most notably "Fast SA" and "Adaptive Simulated Annealing." His web site is

`http://www.ingber.com/`.

[1] The web is an everchanging environment. The authors endeavor to maintain up-to-date links to resources of this section at the book website.

An implementation of simulated annealing to solve a traveling salesman problem can be found in section 10.9 of *Numerical Recipes* (second edition) by Press et al. The software may be downloaded for a fee from:

`http://www.nr.com/.`

Several commercial software products contain simulated annealing modules for solving domain-specific problems in computational science and engineering. Common application areas include circuit design optimization, geologic ground water modeling, and petroleum prospecting. One example, whose forerunners have provided several of the studies cited in this book, is the ISIS package produced by the Danish firm Ødegaard (`www.oedegaard.com`), which is specifically written for the oil industry.

17.1.2 The Methods of This Book

Only one open-source distribution of the methods of this text is currently available. The EBSA (ensemble-based simulated annealing) software package was originally implemented in C and distributed in binary form from the San Diego Supercomputer Center by one of the authors (Frost). It has been entirely rewritten for Fortran 90, Java, and Matlab. The software is composed of a unified set of tools that can be used to construct a simulated annealing application. Interested readers can download it from the authors' web site given above.

17.1.3 Software Libraries

Readers developing their own software solutions should avoid writing fundamental mathematical routines from scratch. Source- or binary-compatible libraries are available for nearly all choices of languages and platforms. For most scientific applications, the NETLIB site is indispensable:

`http://www.netlib.org/.`

In particular, if you need linear algebra the best known source is LAPACK:

`http://www.netlib.org/lapack/.`

For readers using the eigenmethods of this text or spectral methods in general, the ARPACK implementation of the recursive arnoldi methods is recommended:

`http://www.caam.rice.edu/software/ARPACK/.`

Java has limited resources and performance capabilities for scientific computing. This is expected to improve since Java is a very strongly typed language. Hopefully, true source-to-binary compilers will be available for Java by the date this book is published.

Developers need to look to familiar sources for numerical libraries in Java. The numerical zealots at the University of Kentucky have produced a source-to-source translation of LAPACK for Java—see JLAPACK at the LAPACK website above.

Visual Numerics, a recently acquired division of IMSL (see `www.imsl.com`) has a library of numerical routines available for Java:

`http://www.vni.com/products/wpd/jnl/.`

Another vendor, DRA Systems, has produced the OR-Objects package, primarily for use in operations research applications. It contains many useful numerical computing routines in Java but is available only in byte-code format:

`http://opsresearch.com/OR-Objects/.`

Finally, readers should be aware of a robust package for numerical computing in Java from CERN, the Colt package:

`http://tilde-hoschek.home.cern.ch/~hoschek/colt/.`

Colt contains many useful data structure classes (e.g., dynamic size arrays), which are also used by the libraries of numerical routines. Colt is general enough to form the foundation of a numerical application.

17.2 Energy Landscapes Database

The authors intend to maintain a database of landscape structure and structural measures for problems discussed in this text. Collaborators in this effort are welcome. As argued in the other chapters of this book, a broad set of structural measures could lead to better classification of hard optimization problems and possibly an answer to the question posed in Chapter 16. To achieve this goal, information is needed regarding structural parameters of different problems and how they correlate to the efficacy of various algorithmic improvements.

Bibliography

[Aarts89] E. Aarts and H. Korst. *Simulated Annealing and Boltzmann Machines.* John Wiley and Sons, New York, 1989.

[AHM+88] B. Andresen, K. H. Hoffmann, K. Mosegaard, J. Nulton, J. M. Pedersen, and P. Salamon. On lumped models for thermodynamic properties of simulated annealing problems. *Journal de Physique, France*, 49:1485–1492, 1988.

[Aze92] R. Azencott. *Simulated Annealing: Parallelization Techniques.* Wiley-Interscience, New York, 1992.

[Ban94] P. Banerjee. *Parallel Algorithms for VLSI Computer-Aided Design.* Prentice-Hall, Englewood Cliffs, NJ, 1994.

[BB88] R. Brouwer and P. Banerjee. A parallel simulated annealing algorithm for channel routing on a hypercube multiprocessor. In *Proceedings of the International Conference on Computer Design*, 1988, pp. 4–7.

[BBK+96] K. D. Ball, R. S. Berry, R. E. Kunz, F. Y. Li, A. Proykova, and D. J. Wales. From topographics to dynamics on multidimensional potential energy surfaces of atomic clusters. *Science*, 271:259–272, 1996.

[Ber93] R. Stephen Berry. Potential surfaces and dynamics: What clusters tell us. *Chemical Reviews*, 93:2379–2394, 1993.

[BH97] K. Binder and D. W. Heermann. *Monte Carlo Simulation in Statistical Mechanics.* Springer Series in Solid State Sciences, Springer-Verlag, New York, 1997.

[Bin79] K. Binder. *Monte Carlo Methods in Statistical Physics.* Springer-Verlag, Berlin, 1979.

[Bin86] K. Binder. *Monte Carlo Methods in Statistical Physics.* Springer-Verlag, Berlin, 1986.

[BKL75] A. B. Bortz, M. H. Kalos, and J. L. Lebowitz. A new algorithm for Monte Carlo simulation of Ising spin systems. *Journal of Computational Physics*, 17:10–18, 1975.

[BR58] C. K. Burke and M. Rosenblatt. A Markovian function of a Markov matrix. *Annals of Mathematical Statistics*, 29:1112–1122, 1958.

[Bra66] L. Brand. *Differential and Difference Equations*. John Wiley and Sons, New York, 1966.

[Bri62] L. Brillouin. *Science and Information Theory*. Academic Press, New York, 1962.

[Cam85] I. A. Campbell. Random walks on a closed loop and spin glass relaxation. *Journal de Physique Lettres*, 46:L1159–L1162, 1985.

[Čer85] V. Černý. Thermodynamical approach to the traveling salesman problem: An efficient simulation algorithm. *Journal of Optimization Theory and its Applications*, 45:41–55, 1985.

[CLRS2001] T. Cormen, C. Leiserson, R. Rivest, and C. Stein. *Introduction to Algorithms* (2nd edition). MIT Press, Cambridge, MA , 2001.

[Con80] W. Conley. *Computer Optimization Techniques*. Petrocelli Books, New York, 1980.

[Dall00] J. Dall. *Searching Complex State Spaces with Extremal Optimization and Other Stochastic Techniques*. M.S. thesis, University of Southern Denmark, 2000.

[Dall01] J. Dall and P. Sibani. Faster Monte Carlo simulations at low temperatures. The waiting time method. *Computer Physics Communications*, 141:260–267, 2001.

[DS90] G. Dueck and T. Scheuer. Threshold accepting: A general purpose optimization algorithm appearing superior to simulated annealing. *Journal of Computational Physics*, 90:161–175, 1990.

[DW96] J. P. K. Doye and D. J. Wales. The structure and stability of atomic liquids: From clusters to bulk. *Science*, 271:484–487, 1996.

[Fel66a] W. Feller. *An Introduction to Probability Theory and Its Applications*. Vol. I. John Wiley and Sons, New York, 1966.

[Fel66b] W. Feller. *An Introduction to Probability Theory and Its Applications*. Vol. II. John Wiley and Sons, New York, 1966.

[FH91] K. H. Fischer and J. A. Hertz. *Spin Glasses*. Cambridge University Press, Cambridge, UK, 1991.

[FH00a] A. Franz and K. H. Hoffmann. Optimal annealing schedules for a modified Tsallis statistics. *Journal of Computational Physics*, 176:196–204, 2002.

[FH00b] A. Franz and K. H. Hoffmann. Threshold accepting as limit case for a modified Tsallis statistics. *Applied Mathematics Letters*, to appear.

[FHS01] A. Franz, K. H. Hoffmann, and P. Salamon. Best possible strategy for finding ground states. *Physical Review Letters*, 86:5219–5222, 2001.

[Fra83] J. Franklin. Mathematical Methods of Economics. *American Mathematical Monthly*, 90:229–244, 1983.

[FS88] A. M. Ferrenberg and R. H. Swendsen. New Monte Carlo Technique for Studying Phase Transitions. *Phys. Review Letters*, 61:2635–2638, 1988.

[FS89] A. M. Ferrenberg and R. H. Swendsen. Optimized Monte Carlo Data Analysis. *Physical Review Letters*, 63:1195–1198, 1989.

[Gan59] F. R. Gantmacher. *Application of the Theory of Matrices*. Wiley-Interscience, New York, 1959.

[Geh97] W. Gehring. *Correlation Structure of Landscapes of NP-Complete Optimization Problems at Finite Temperature*. M.S. thesis, San Diego State University, 1997.

[GG84] S. Geman and D. Geman. Stochastic Relaxation, Gibbs Distributions, and the Bayesian Restoration of Images. In *Proceedings of the Sixth IEEE Pattern Analysis and Machine Intelligence*, 1984, pp. 721–741.

[GJ79] M. R. Garey and D. S. Johnson. *Computers and Intractability: A Guide to the Theory of NP Completeness*. W. H. Freeman, New York, 1979.

[Gold89] D. Goldberg. *Genetic Algorithms in Search, Optimization, and Machine Learning*. Addison-Wesley, Reading, MA, 1989.

[Gre80] D. M. Greig. *Optimisation*. Longman, London, 1980.

[GS84] J. W. Greene and K. J. Supowit. Simulated annealing without rejected moves. In *Proceedings of the ICCD*, 1984, p. 646.

[GSL86a] G. S. Grest, C. M. Soukoulis, and K. Levin. Comparative Monte Carlo and mean-field studies of random-field Ising systems. *Physical Review B*, 33:7659–7674, 1986.

[GSL86b] G. S. Grest, C. M. Soukoulis, and K. Levin. Cooling rate dependence for the spin glass ground state energy: Implications for optimization by simulated annealing. *Physical Review Letters*, 56:1148–1151, 1986.

[GWF97] R. S. Giordano, M. D. Weir, and W. P. Fox. *A First Course in Mathematical Modeling*. Brooks/Cole Publishing, Pacific Grove, CA, 1997.

[GWH85] S. Grossmann, F. Wegner, and K. H. Hoffmann. Anomalous diffusion on a selfsimilar hierarchical structure. *Journal de Physique Lettres*, 46:575–583, 1985.

[Haj88] B. Hajek. Cooling Schedules for Optimal Annealing. *Mathematics of Operations Research*, 13:311–329, 1988.

[HKP91] J. A. Hertz, A. Krogh, and R. Palmer. *Introduction to the Theory of Neural Computation*. Addison-Wesley, Redwood City, CA, 1991.

[HM97] G. A. Huber and J. A. McCammon. Weighted-ensemble simulated annealing: Faster optimization on hierarchical energy surfaces. *Physical Review E*, 55:4822–4825, 1997.

[Hol75] J. H. Holland. *Adaptation in Natural and Artificial Systems: An Introductory Analysis with Applications to Biology, Control, and Artificial Intelligence.* University of Michigan Press, Ann Arbor, 1975.

[HRS98] R. Haupt and S. Haupt. *Practical Genetic Algorithms.* John Wiley and Sons, New York, 1998.

[HRSV86] M. D. Huang, F. Romeo, and A. Sangiovanni-Vincentelli. An efficient general cooling schedule for simulated annealing. In *Proceedings of the International Conference on Computer Aided Design*, 1986, p. 381.

[HS90] K. H. Hoffmann and P. Salamon. The optimal simulated annealing schedule for a simple model. *Journal of Physics A*, 23:3511, 1990.

[HSPS90] K. H. Hoffmann, P. Sibani, J. M. Pedersen, and P. Salamon. Optimal ensemble size for parallel implementations of simulated annealing. *Applied Mathematics Letters*, 3:53–56, 1990.

[HWdGH91] K. H. Hoffmann, D. Würtz, C. de Groot, and M. Hanf. Concepts in optimizing simulated annealing schedules: An adaptive approach for parallel and vector machines. In *Parallel and Distributed Optimization*, M. Grauer and D. B. Pressmar, editors, Springer Verlag, Heidelberg, 1991.

[Huang87] K. Huang. *Statistical Mechanics.* John Wiley and Sons, New York, 1987.

[Ing89] L. Ingber. Very fast simulated re-annealing. *Mathematical and Computer Modelling*, 12:967–973, 1989.

[JAMS89] D. S. Johnson, C. R. Aragon, L. A. McGeogh, and C. Schevon. Optimization by simulated annealing—an experimental evaluation. 1. Graph partitioning. *Operations Research*, 37:865–892, 1989.

[JAMS91] D. S. Johnson, C. R. Aragon, L. A. McGeogh, and C. Schevon. Optimization by simulated annealing—an experimental evaluation. 2. Graph-coloring and number partitioning. *Operations Research*, 39:378–406, 1991.

[Jay83] E. T. Jaynes. *Papers on Probability, Statistics and Statistical Physics.* D. Reidel Publishing, Dordrecht, Boston, 1983.

[JM97] D. S. Johnson and L. A. McGeogh. The traveling salesman problem: A case study. In *Local Search in Combinatorial Optimization*, E. Aarts and J. K. Lenstra, editors, Wiley-Interscience, New York, 1997, pp. 215–310.

[JMP88] M. O. Jakobsen, K. Mosegaard, and J. M. Pedersen. Global model optimization in reflection seismology by simulated annealing. In *Model Optimization in Exploration Geophysics* II, A. Vogel, editor, Friedr. Vieweg & Son, Braunschweig, 1988, pp. 361–381.

[JMS96] B. H. Jacobsen, K. Mosegaard, and P. Sibani. *Inverse Methods, Interdisciplinary Elements of Methodology, Computation, and Applications.* Springer-Verlag, New York, 1996.

[JP94] J. S. Jørgensen and J. B. Pedersen. Calculation of the variability of model parameters. *Chemometrics and Intelligent Laboratory Systems*, 22:25–35, 1994.

[Kam92] N. G. Van Kampen. *Stochastic Processes in Physics and Chemistry*. North-Holland, Dordrecht, 1992.

[KJV83] S. Kirkpatrick, C. D. Gelatt, Jr., and M. P. Vecchi. Optimization by simulated annealing. *Science*, 220:671–680, 1983.

[Koza99] J. Koza. *Genetic Programming* III: *Automatic Programming and Automatic Circuit Synthesis*. Morgan Kaufmann, San Francisco, 1999.

[Kra40] H. A. Kramers. Brownian motion in a field of force and the diffusion model of chemical reactions. *Physica*, 7:284–304, 1940.

[KS60] J. G. Kemeny and J. L. Snell. *Finite Markov Chains*. Van Nostrand, Princeton, NJ, 1960.

[Kh49] A. I. Khinchin. *Mathematical Foundations of Statistical Mechanics*. Dover, New York, 1949.

[Kul68] S. Kullback. *Information Theory and Statistics*. Dover, New York, 1968.

[LB00] D. P. Landau and K. Binder. *A Guide to Monte Carlo Simulations in Statistical Physics*. Cambridge University Press, Cambridge, UK, 2000.

[LD87] J. Lam and J. M. Delosme. An Adaptive Annealing Schedule. Report 8608, Yale University, Department of Electrical Engineering, New Haven, CT, 1987.

[LD99] R. Leary and J. Doye. New tetrahedral global minimum for the 98-atom Lennard-Jones cluster. *Physical Review E*, 60:R6320–R6322, 1999.

[Lue84] D. G. Luenberger. *Linear and Nonlinear Programming*. Addison-Wesley, New York, 1984.

[McQ00] D. A. McQuarrie. *Statistical Mechanics*. University Science Books, Sausalito, CA, 2000.

[MF90] P. Moscato and J. F. Fontanari. Stochastic versus deterministic update in simulated annealing. *Physics Letters A*, 146:204–208, 1990.

[MNDS+97] A. Möbius, A. Neklioudov, A. Díaz-Sánchez, K. H. Hoffmann, A. Fachat, and M. Schreiber. Optimization by Thermal Cycling. *Physical Review Letters*, 79:4297–4301, 1997.

[Morey98] C. Morey, J. A. Scales, and E. S. Van Vleck. A feedback algorithm for determining search parameters for Monte Carlo optimization. *Journal of Computational Physics*, 146:263–281, 1998.

[MR81] E. W. Montroll and H. Reiss. Phase transition versus disorder: A criterion derived from a two-dimensional dynamic ferromagnetic model. *Proceedings of the National Academy of Sciences*, 78:2659–2663, 1981.

[MRR+53] N. Metropolis, A. W. Rosenbluth, M. N. Rosenbluth, A. H. Teller, and E. Teller. Equation of state calculations by fast computing machines. *Journal of Chemical Physics*, 21:1087–1092, 1953.

[MRX92] R. S. Maier, J. B. Rosen, and G. L. Xue. A discrete-continuous algorithm for molecular energy minimization. In *Proceedings of the ACM International Conference on Supercomputing*, Washington, D.C., 1992, pp. 778–786.

[MT95] K. Mosegaard and A. Tarantola. Monte Carlo sampling of solutions to inverse problems. *Geophysical Research B*, 100:12431–12447, 1995.

[MV91] K. Mosegaard and P. D. Vestergaard. A simulated annealing approach to seismic model optimization with sparse prior information. *Geophysical Prospecting*, 39:599–611, 1991.

[NA98] Y. Nourani and B. Andresen. A comparison of simulated annealing cooling strategies. *Journal of Physics A*, 31:8373, 1998.

[NA99] Y. Nourani and B. Andresen. Exploration of NP-hard enumeration problems by simulated annealing—the spectrum values of permanents. *Theoretical Computer Science*, 215:51, 1999.

[NS88] J. Nulton and P. Salamon. Statistical mechanics of combinatorial optimization. *Physical Review A*, 37:1351, 1988.

[NSAA85] J. Nulton, P. Salamon, B. Andresen, and Q. Anmin. Quasistatic processes as step equilibrations. *Journal of Chemical Physics*, 83:334, 1985.

[OS85] A. T. Ogielski and D. L. Stein. Dynamics on ultrametric spaces. *Physical Review Letters*, 55:1634–1637, 1985.

[Otten89] R. Otten and L. Van Ginneken. *The Annealing Algorithm*. Kluwer, Boston, 1989.

[Pal82] R. G. Palmer. Broken ergodicity. *Advances in Physics*, 31:669–735, 1982.

[Ped90] J. M. Pedersen. *Simulated Annealing and Finite-Time Thermodynamics*. Ph.D. dissertation, University of Copenhagen, Physics Institute, 1990.

[Pen94] T. J. P. Penna. Traveling salesman problem and Tsallis statistics. *Physical Review E*, 51:R1–R3, 1994.

[PFTV81] W. H. Press, B. P. Flannery, S. A. Teukolsky, and W. T. Vetterling. *Numerical Recipes*. Cambridge University Press, Cambridge, UK, 1981.

[RPS91] G. Ruppeiner, J. M. Pedersen, and P. Salamon. Ensemble approach to simulated annealing. *Journal de Physique* I, 1:455–470, 1991.

[Rud76] W. Rudin. *Principles of Mathematical Analysis*. McGraw-Hill, New York, 1976.

[Rup88] G. Ruppeiner. Implementation of an adaptive constant thermodynamic speed simulated annealing schedule. *Nuclear Physics B*, 5A:116–121, 1988.

[SA94] D. Stauffer and A. Aharony. *Introduction to Percolation Theory*. Taylor and Francis, London, 1994.

[SB83] P. Salamon and R. S. Berry. Thermodynamic length and dissipated availability. *Physical Review Letters*, 51:1127–1130, 1983.

[Sch97] J. C. Schön. Preferential trapping on energy landscapes in regions containing deep-lying minima: The reason for the success of simulated annealing? *Journal of Physics A*, 30:2367–2389, 1997.

[SH87] H. Szu and R. Hartley. Fast simulated annealing. *Physics Letters A*, 122:157, 1987.

[SH89] P. Sibani and K. H. Hoffmann. Hierarchical models for aging and relaxation in spin glasses. *Physical Review Letters*, 63:2853–2856, 1989.

[Sha49] C. Shannon. *The Mathematical Theory of Communication*. University of Illinois Press, Urbana, 1949.

[SHHN88] P. Salamon, K. H. Hoffmann, J. Harland, and J. D. Nulton. *An Information Theoretic Bound on the Performance of Simulated Annealing Algorithms*. Technical Report 88–1, San Diego State University, Department of Mathematical Sciences, 1988.

[Sib87] P. Sibani and K. H. Hoffmann. Random walks on Cayley trees: Temperature induced transience-recurrence transition, small exponents and logarithmic relaxation. *Europhysics Lett.*, 4:967–972, 1987.

[Sib98] P. Sibani. Local state space geometry and thermal relaxation in complex landscapes: The spin-glass case. *Physica A*, 258:249–262, 1998.

[SMS98] J. Schneider, I. Morgenstern, and J. M. Singer. Bouncing towards the optimum — improving the results of Monte Carlo optimization algorithms. *Physical Review E*, 58:5085–5095, 1998.

[SNR+88] P. Salamon, J. D. Nulton, J. Robinson, J. M. Pedersen, G. Ruppeiner, and L. Liao. Simulated annealing with constant thermodynamic speed. *Computer Physics Communications*, 49:423–428, 1988.

[SPHS90] P. Sibani, J. M. Pedersen, K. H. Hoffmann, and P. Salamon. Monte Carlo dynamics of optimization problems, a scaling description. *Physical Review A*, 42:7080–7086, 1990.

[SS92] P. F. Stadler and W. Schnabl. The landscape of the traveling salesman problem. *Physics Letters A*, 161:337–344, 1992.

[SS94] P. Sibani and P. Schriver. Phase-structure and low-temperature dynamics of short range Ising spin glasses. *Physical Review B*, 49:6667–6671, 1994.

[SS98] J. C. Schön and P. Sibani. Properties of the energy landscape of network models for covalent glasses. *Journal of Physics. A*, 31:8165–8178, 1998.

[SS00] J. C. Schön and P. Sibani. Energy and entropy of metastable states in glassy systems. *Europhysics Letters*, 49:196–202, 2000.

[SSA95] P. Sibani, M. Schmidt, and P. Alstrøm. Fitness optimization and decay of the extinction rate through biological evolution. *Physical Review Letters*, 75:2055–2058, 1995.

[SSSA93] P. Sibani, C. Schön, P. Salamon, and J.-O. Andersson. Emergent hierarchical structures in complex system dynamics. *Europhysics Letters*, 22:479–485, 1993.

[SSV85] C. Sechen and A. Sangiovanni-Vincentelli. The timberwolf placement and routing package. *IEEE Journal of Solid State Circuits*, SC-20:510–522, 1985.

[Sta92] P. F. Stadler. Correlation in landscapes of combinatorial optimization problems. *Europhysics Letters*, 20:479–482, 1992.

[Sun95] A. C. Sun and W. D. Seider. Homotopy-continuation algorithm for global optimization. In *Recent Advances in Global Optimization*, Princeton University Press, Princeton, NJ, 1992, pp. 561–592.

[SvdPS99] P. Sibani, R. van der Pas, and J. C. Schön. The lid method for exhaustive exploration of metastable states of complex systems. *Computer Physics Communications*, 116:17–27, 1999.

[SW88] F. H. Stillinger and T. A. Weber. Nonlinear optimization simplified by hypersurface deformation. *Journal of Statistical Physics*, 52:1429–1445, 1988.

[TH95] R. Tafelmayer and K. H. Hoffmann. Scaling features in complex optimization problems. *Computer Physics Communications*, 86:81–90, 1995.

[TS96] C. Tsallis and D. A. Stariolo. Generalized simulated annealing. *Physica A*, 233: 395–406, 1996.

[TY86] L. E. Thomas and Z. Yin. Approach to equilibrium for random walks on graphs and for stochastic infinite particle processes. *Journal of Mathematical Physics*, 27:2475–2477, 1986.

[UHS95] C. Uhlig, K. H. Hoffmann, and P. Sibani. Relaxation in self similar hierarchies. *Zeitschrift für Physik B*, 96:409–416, 1995.

[Ven92] V. R. Venkataraman. *The Neighbourhood Structure in the Graph Partitioning Problem*. M.S. thesis, San Diego State University, 1992.

[vLA87] P. J. M. van Laarhoven and E. H. L. Aarts. *Simulated Annealing*. D. Reidel Publishing, Dordrecht, 1987.

[VM91] P. D. Vestergaard and K. Mosegaard. Inversion of post-stack seismic data using simulated annealing. *Geophysical Prospecting*, 39:613–624, 1991.

[W90] E. Weinberg. Correlated and uncorrelated fitness landscapes and how to tell the difference. *Biological Cybernetics*, 63:325–336, 1990.

[Whi84] S. R. White. Concepts of scale in simulated annealing. In *IEEE Proceedings of the* 1984 *International Conference on Computer Design*, 1984, pp. 646–651.

[WS99] D. J. Wales and H. A. Scheraga. Chemistry: Global optimization of clusters, crystals and biomolecules. *Science*, 285:1368–1372, 1999.

[YBS91] J.-Y. Yi, J. Bernholc, and P. Salamon. Simulated annealing strategies for molecular dynamics. *Computer Physics Communications*, 66:177–180, 1991.

[ZS92] T. Zimmermann and P. Salamon. The demon algorithm. *International Journal of Computer Mathematics*, 42:21–32, 1992.

Index